Signal Integrity

Samuel H. Russ

Signal Integrity

Applied Electromagnetics and Professional Practice

Second Edition

 Springer

Samuel H. Russ
University of South Alabama
Mobile, AL, USA

ISBN 978-3-030-86929-8   ISBN 978-3-030-86927-4   (eBook)
https://doi.org/10.1007/978-3-030-86927-4

This Springer imprint is published by the registered company Springer Nature Switzerland AG
The registered company address is: Gewerbestrasse 11, 6330 Cham, Switzerland

Introduction

Why Study This Subject?

It is my hope that this book will help the reader understand *applied electromagnetics*. The book will be getting into some fairly complicated and very abstract topics, but remember – the emphasis will be on the practical or "applied" aspects of the subject.

Have you ever wondered why the flight attendant tells you to turn off your cell phone at takeoff? Have you ever wondered how your hard drive talks to your computer at 3 GHz per second over a very thin cable? How did the company that makes your computer's processor get it to work at 3 GHz?

All of these daily examples are part of the world of *signal integrity*. Signal integrity is the science (and art) of designing systems that carry signals intact over distance and that do not interfere with each other. The subject lies at the intersection of electromagnetics and computer engineering, and so understanding it is vital for computer engineers in the Gigahertz era.

Why This Book?

One may wonder whether a textbook like this is worth the cost. In the case of this book, it turns out there are two good reasons. First, this subject is extremely important and rarely taught in college. So, this book is not only what students will use to take this class, it can also become a permanent part of a technical bookshelf, and might even bail you out if you get stuck on a signal integrity issue. This book is designed specifically for that purpose – the technical bailout. Second, the book has plenty of examples and some real-world testimonials based on my own experiences.

Why Are We Here, and Where Is "Here"?

The twin revolutions of computing and communications have achieved the dream of a world with nearly instantaneous access to all corners of the globe and with a seemingly limitless body of knowledge at our fingertips. All of this was brought about through the hard work of roughly three generations of electrical and computer engineers who developed the computer chips, interconnections, and communications systems that made it all possible.

As the twin revolutions were unfolding in the 1995 time frame (right about the time the World Wide Web was launched), clock frequencies and data rates nudged up into hundreds of megahertz. (I still remember my whopping 90 MHz Pentium computer.)

At this point, a very crucial change occurred.

It turns out that there are two ways of thinking about a wire. One can use a "lumped" approximation and assume the wire is a zero-ohm resistor with the same voltage at every point, or one can use a "distributed" approximation and assume the wire has inductance and can have different voltages at different points.

When the 100-MHz barrier was crossed (give or take a megahertz or two), wires on computer motherboards stopped being wires (in the lumped sense) and started becoming transmission lines (i.e., distributed wires).

The only way to design systems today that work correctly is to use distributed analysis. This is the traditional domain of electromagnetics experts. Computer engineers literally need to know most of the methods that were used to design microwave systems back in the 1960s. This is where we are today – in a world that must be modeled using distributed analysis.

This book will teach how to convert fluently between frequency, time, and distance. This book will then teach how signals propagate, how they interfere, what can go wrong, and how to fix it. The focus in this book is on the design of real-world systems using physical principles and in cultivating an engineering intuition based on physics and measurements. Along the way, the book also looks at the design of digital systems from a manager's perspective so you can both be a better informed engineer and become a successful manager one day (if you want to).

What makes digital design challenging? I think the challenge lies in three very common ingredients. First, digital designs are almost always cost constrained, especially consumer items like MP3 players. Second, the increasing clock speeds (more accurately, the faster rise and fall times) make the design more complicated because everything has to work up to higher frequencies. Third, there is a considerable time-to-market pressure in the fast-changing world of computer engineering. (Did you know that Motorola lost to Intel when IBM designed the first PC because the Motorola processor was 6 months late? A 6-month slip in schedule forever changed the microprocessor landscape!) You have to get the design working and into mass production quickly, without much time to make or fix mistakes.

So, armed with this book, students and professionals can embark on successful digital designs.

Contents

Chapter 1
The Basics – Charge, Energy, Time, and Distance

Background and Objectives

The concept of signal integrity begins with the basics – charge, energy, time, and distance. Electrical and computer engineering students already know a lot about these basic building blocks of electrical engineering, but a review is in order so that the aspects relevant to signal integrity can be emphasized. When this chapter is finished, you should be able to:

- Recall the relationships between charge, current, power, and energy
- Understand the basic concepts of E fields and B fields
- Express the speed of light both in terms of distance per time and in terms of time per distance
- Calculate the speed of propagation in terms of dielectric constant
- Define rise time, calculate the length of a rising edge, and define and calculate the critical length
- Calculate the effect of adding a circuit with a nonzero rise time
- Define and calculate the knee frequency of a digital signal

Getting Back to Basics

The review begins with some concepts that most electrical engineers think they understand, but might not. For starters, what, exactly, is a volt?

I specifically remember when a professor in one of my graduate classes asked that question, and at the time I did not have a really good answer. It turns out to have an embarrassingly simple answer: *amount of energy per electron*. Think about the units – 1 Volt equals 1 Joule per Coulomb. Of course, there are some unit conversions involved because energy is measured in Joules and there are 6.2×10^{18} electrons is in a Coulomb of charge, but voltage boils down to energy per electron.

© Springer Nature Switzerland AG 2022
S. H. Russ, *Signal Integrity*, https://doi.org/10.1007/978-3-030-86927-4_1

There are other, more sophisticated answers to the question "what is a volt?", but I find that this answer is the simplest and leads to an intuition about the actual physical processes involved in a circuit.

This intuitive definition of a volt leads to other important concepts, such as the fact that a voltage drop means that electrical energy is being lost, or a voltage gain means that some other source of energy is being converted to electrical energy. Energy is conserved, so a drop in voltage means that electrical energy is being lost; there is no other explanation.

The next basic concept is current. This is very easy to grasp: Current boils down to the *amount of charge flowing past a point (or through a surface) per second.* It is a rate and so has units of Coulombs per second.

And, of course, the product of current (Coulombs per second) and voltage (Joules per Coulomb) is power (Joules per second). The product of power (Joules per second) and time (seconds) is energy (Joules). (These are examples of how units' analysis can be a helpful way to remember basic concepts.)

The next two concepts are much more abstract.

The first is the concept of an *E field*. A helpful way to remember E fields is to remember that the unit of measurement of an E field is volts per meter. If there is a difference in voltage between two points, then there is an E field between them. What happens next depends on the material between the two points. That is, what happens next depends on the material that contains the E field.

If the material containing the E field can conduct electricity (i.e., can carry freely moving electrons), then current flows as electrons move from a state of high energy (the higher-voltage point) to a state of low energy (the lower-voltage point). In such a case, electrons are behaving like little balls, rolling "downhill."

If the material containing the E field cannot conduct electricity, then there is a static E field between the two points. There is more to the story – the E field can actually store energy and convey information, as we will see – but that is the basic idea.

The E field is often misunderstood. Many of the properties that the average engineer associates with electrons are actually due to E fields. For example, data flowing over a Serial ATA (SATA) cable to a hard disk drive is actually traveling *in the E field between the conductors in the wire*, and not on the conductors themselves. If you have any doubt on this point, ask yourself how your next selfie gets from your phone to the cell phone antenna. *There is no wire.* But there is an E field between the antenna of the phone and the antenna of the cell tower. Moving electrons are necessary – they are the most practical way to carry real energy – but E fields tend to be used to carry information because, as we will see, E fields (more specifically, changes to E fields) move really fast.

The second is the concept of a B field. The B field is a magnetic field due to the fact that moving electrons create a magnetic field. That is, one way to create a B field is to move electrons through a region. The linkage between moving electrons and B fields occurs because of relativity, it turns out, but the point is that more moving electrons, generally speaking, results in more of a magnetic field, and so the B field is tied to current. Like its counterpart E field, it can store energy.

The E and the B fields interact in complicated ways. (Electromagnetics courses cover this in much more detail.) One important aspect of these fields is that they surround the conductors that we typically think of as carrying signals (i.e., they lie outside of the conductors), and so they can cause effects remotely. Many, if not most, signal integrity issues are a result of this "spooky action at a distance" property of these fields. It creates a world in which rapidly changing digital signals become "noisy neighbors." (It also has advantages, like radio, but this book is not about building radios.)

Consequences of the Basics

So these concepts can be connected. Following are some examples.

Question 1:   What defines the speed of propagation of an electromagnetic wave (i.e., the speed of a changing E field) in an insulator?

Answer: The speed of propagation is a function of the medium's permittivity (which determines its capacitance, in turn related to E field) and the medium's permeability (its inductance, in turn related to B field). The "medium" is the electrical insulator that the wave is traveling in. The medium from the cell antenna to your phone is air. The medium of a trace on your PC motherboard is the fiberglass circuit-board material between the signal-carrying conductor and the ground plane. In both cases, the medium is the insulator that carries the signal energy. (Note that, if the medium was a conductor, the E field would turn into the simple flow of current.)

Question 2:   True or false: Electric current follows the path of least resistance.

Answer: FALSE! It follows the path of least IMPEDANCE. (Make sure you remember this!!!) What is the difference? As you know, resistance is DC and impedance is AC. Inside the circuit boards of busy digital systems, essentially everything is AC. (Technically speaking, everything that matters are a transient, but transients are an AC phenomenon, not a DC phenomenon.) Even the power lines on a circuit board are AC because they have to supply large amounts of instantaneous current when an output switches. That is, even the power lines on a circuit board have transients and therefore must be analyzed as AC signals. A DC signal is one that never changes, which is a signal that carries no information. Almost all "signals" are AC, and so take the path of least impedance.

Question 3:   What does a resistor look like electrically?

Answer: In basic circuits class, we all learned that a resistor is, well, a resistor! (We will study resistance in detail again in Chap. 4). But a more complete model of a resistor has to take into account the fact that the physical package (that the resistor comes in) has inductance and the presence of the resistor near a ground plane creates capacitance. So, a resistor looks like a resistor in series with a small inductance, and the resistor-inductor pair is connected to ground through a small capacitor.

This is the point: Circuit elements (resistors, wires, etc.) are not what they appear at very high frequencies. In the case of the resistor, to continue the example, the frequency where the inductor becomes significant is the point at which the magnitude of the impedance of the inductor becomes a significant fraction of the value of the resistor. At even higher frequencies, the impedance of the inductor dominates and the resistor behaves like an inductor.

Distance, Time, Speed, and c

Most are familiar with the concepts of time and distance and their ratio, speed. It turns out that the most basic units of the three are time and speed, and distance is derived from the other two.

One second is a duration corresponding to 9,192,631,770 periods of a type of radio frequency energy emitted and absorbed by cesium atoms. In other words, scientists define the cesium-atom frequency as exactly 9.19263177 GHz and use it to define one second.

The speed of light in a vacuum is an often-used physical constant of the universe. It is denoted by c (the first letter of the Latin word for "speed"). Maxwell's equations make clear that light (and other electromagnetic waves) travel at a certain, fixed speed. The speed c is found by $c = 1/\sqrt{\mu_0 \varepsilon_0}$, where μ_0 is the permeability of free space and ε_0 is the permittivity of free space. "Free space" is the same thing as "in a vacuum" and will be used interchangeably. From an electrical standpoint, air is very close to a vacuum, and so radio waves travel through the atmosphere at a speed very close to c.

As noted above, μ_0, the permeability of free space, is associated with magnetic properties, inductance, and the B field. ε_0, is the permittivity of free space, is associated with electric-field properties, capacitance, and the E field. Moving electromagnetic waves are the result of interactions between the E and B fields, and so their speed is a function of both fields. At any rate, light travels in a vacuum at a fixed speed (as far as we know) everywhere in the universe. In fact, the direction in which one is traveling does not matter; one always measures the speed of light to be the same. (This property is where "relativity" gets its name.)

The definition of distance is actually derived from the definition of time and the actual speed of light. The definition of a meter is the length traversed by light in 1/299,792,458 of a second. So, you have to know what a "second" is, and what the speed of light is, in order to know what a meter is. One consequence of this definition is that the speed of light is, by definition, 299,792,458 meters per second, or almost exactly 300,000 km/s.

Our entire system of measurement is based on two basic facts, the frequency of radio emissions from cesium and the speed of light in a vacuum.

The speed of light is so high that it has little physical or intuitive meaning, so a good way to remember it is this: How long does light travel in one nanosecond? The math is pretty simple, 299,792,458 divided by 1 billion, or almost exactly 0.3 meters. In old-fashioned English units, it turns out to be 11.8 inches, or almost exactly one foot. The speed of light is 0.3 m, 30 cm, or 1 foot per nanosecond. This leads to an extremely important mental picture: If you hold your hands 30 cm (or 1 foot) apart, that is how far light travels in one nanosecond. This is roughly how fast, for example, a signal travels down a fiber-optic cable or a radio wave travels through air. On a circuit board, where things travel at roughly half the speed of light (more on why this is later), your hands are *two nanoseconds apart*.

This ability – the ability to transition between speed, time, and distance – is an important skill in signal integrity. Circuit boards are several centimeters in size and the signals on them travel in a few nanoseconds from one point to another. The ability to convert between size and time requires knowledge of speed, the ratio of the two.

It is often convenient to use the reciprocal of the speed of light. Expressed as a reciprocal, the speed of light is 84.7 ps/inch or 33.36 ps/cm. This reciprocal version may seem odd, but it lets one convert a physical distance to a time delay. Again, this maps the relatively abstract concept of the speed of light to a much more usable form, such as the amount of time it takes for a signal to get from one point to another.

Example 1.1: Speed, Time, and Distance

Consider a signal traveling on a circuit board at ½ of the speed of light. The trace that the signal travels along is 10 cm long. How long does it take to get from one end to the other?

The speed of light is 33.36 ps/cm.

The signal on the circuit board travels at half the speed of light, so what is the speed of the signal? The ratio is clearly ½, but do you multiply 33.36 times 2 or divide 33.36 by 2? Here is a helpful way to think of it: The signal is traveling *slower* than the speed of light, so the time it takes to travel a unit of distance goes *up*. Hence, 33.36 ps/cm is multiplied by 2.

Since the signal travels at half that speed, it travels at 66.72 ps/cm. The trace is 10 cm long, so it takes $66.72 \times 10 = 667$ ps, or about $2/3^{rd}$ of a ns. The signal travels 10 cm in about $2/3^{rd}$ of a ns.

The Effect of Dielectrics

So enough of this talk about "in a vacuum." What about here on earth?

Well, the first thing to note is that almost all materials, such as air, paper, and fiberglass, have a permeability close to that of free space. The most common exceptions are conductors that are ferromagnetic (like iron, cobalt, and nickel). In other words, most materials have a permeability close to free space because they are not very magnetic. What determines the speed of propagation turns out to be the permittivity.

To discuss permittivity, it becomes convenient to express a material's permittivity (also called its *dielectric constant*) relative to the permittivity of free space. We will refer to *relative permittivity* as ε_r and so, for example, the relative permittivity of FR-4 fiberglass (the type used in circuit boards) is 4.2, meaning that its permittivity is 4.2 times that of free space. That also means a capacitor with FR-4 insulator will have 4.2 times the capacitance of the same sized capacitor made with air or vacuum insulator. A convenient aspect of relative permittivity is that one never has to remember the value of the permittivity of free space. (Permittivity, in turn, controls the amount of capacitance a volume of the material can contain. We will learn about capacitance later in the book.)

So, what is the speed of "light" (i.e., the speed of electromagnetic waves or the speed of a changing E field) inside a circuit board? $speed = \frac{1}{\sqrt{\mu\varepsilon}} = \frac{1}{\sqrt{\mu_0 4.2\varepsilon_0}} = \frac{c}{\sqrt{4.2}}$.

Noting that the square root of 4.2 is close to 2, it means that electromagnetic waves travel at about $c/2$ inside a circuit board, corresponding to about 6 inches or 0.15 m per nanosecond. Since the most common circuit-board insulator is fiberglass, this explains why the speed of signals on most circuit boards is about $c/2$.

This can be expressed in general. The speed of propagation s in a medium with a relative dielectric constant ε_r is found by

$$s = c/\sqrt{\varepsilon_r} \tag{1.1}$$

This formula assumes that the conductive material is not ferromagnetic.

Example 1.2: Circuit-Board Propagation

An engineer is designing a memory circuit on a circuit board that has $\varepsilon_r = 3$. The circuit is allowed to have a delay of 200 ps. How long can a circuit trace be on the board?

The speed of light is 30 cm/ns. On the circuit board, the speed is $c/\sqrt{3} = 30/\sqrt{3} =$ 17 cm/ns . Since 200 ps is 0.2 ns, the maximum length is 17 cm/ns $\times$ 0.2 ns = 3.4 cm.

Rise Time and Bandlimit

Digital signals are normally only interesting when they change. Assuming that 0 Volts is a logic 0 and, say, 5 Volts is a logic 1, one often thinks of the transition from a logic 0 to a logic 1 as a square wave. In fact, it is not. The rising voltage takes some nonzero time to make the transition because the driving gate has finite drive strength and the circuit being driven has resistance and capacitance.

The time it takes to make the transition from a low voltage to a high voltage is called *rise time*, which we will denote t_r. In this book, we will define the rise time in terms of what is called the *10% to 90% rise time*, the time it takes the voltage to rise from 10% of the distance between a logic 0 and a logic 1 to 90%. For example, if a logic 1 is 5 V, the rise time is the time it takes the voltage to rise from 0.5 to 4.5 V. (We will refer to the voltage value of a logic 1 as "Vdd" or "Vcc.") Rise time is illustrated below in Fig. 1.1.

Fall time is defined similarly (the time to go from 90% to 10%), and the term "rise time" is often used to refer to "rise or fall time" because often the two are (often) symmetric.

Because rise time is a time duration, a faster rise time has a smaller numeric value than a slower rise time. This leads to the odd linguistic property that a "higher" rise time is slower. As much as possible, this text will use the term "faster" rise time to make the situation less ambiguous.

Have you ever considered the fact that a rising edge has a physical length? The voltage rise occurs over a nonzero amount of time, and so it must have (due to the speed of light) a nonzero length. The length of a rise time of duration t_r can be calculated by multiplying the time by the speed, since distance equals time times speed. In other words, a rise time of duration t_r over a dielectric with relative permittivity ε_r has a length

Fig. 1.1 Definition of rise time. The rise time is defined as the interval from the signal being at 10% of a full logic swing to 90%

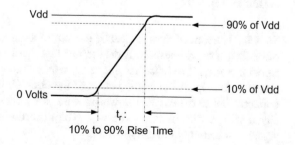

$$\ell = t_r \; c/\sqrt{\varepsilon_r} \tag{1.2}$$

For a rise time measured in ns and a length in cm (and recalling that $c = 30$ cm/ns),

$$\ell = 30 t_r /\sqrt{\varepsilon_r} \tag{1.3}$$

Example 1.3: Length of a Rising Edge
A signal in the memory circuit from the previous example has a rise time of 50 ps. How long is the rising edge?

The rising edge is 50 ps $= 0.05$ ns. $\ell = (30 \times 0.05)/\sqrt{3} = 0.866$ cm.

Note the conversion from ps to ns – this was necessary because the formula was written for a time in ns.

It turns out that the rise time is an extremely important parameter in a digital system. The rising (or falling) edge is the fastest rate of change in the system, and so it represents the maximal frequency content (or *bandlimit*) of the system.

This notion of frequency content is very important in circuit design. Electrical engineering is often performed in the frequency domain, and knowing the bandlimit of a signal enables an engineer to know how demanding a circuit design needs to be. For example, designing for a signal with a bandlimit of 1 MHz is much easier than designing for a signal with a bandlimit of 1 GHz.

The correlation between the rise time and bandlimit may seem murky, but it needs to be unpacked. A rising edge is the fastest signal in any digital system. Signals have to rise (or fall) to their final value before logic gates can operate on that value. Since they are the fastest signals, they have the steepest slopes. In other words, they have the highest dV/dt or the highest volts per second.

As electrical engineers, we are trained to think of frequency as something involving sine waves. This is only a convenience; frequency is more accurately a function of the reciprocal of time, the rate of change of a signal quantity. This distinction is important because rising and falling edges are not sine waves, but the signal that they represent has a bandlimit.

Knee Frequency

So, what is the maximum frequency content of a digital signal? This turns out to be more than a trivia question: It determines how hard you have to work to get good signal integrity. If a signal has frequency content out to 5 GHz, then the wiring that conducts the signal (the combination of the circuit board and cabling) has to work correctly out to 5 GHz. If the wiring does not work correctly at 5 GHz, then the signal will be distorted in some way. The distortion might be acceptable, but the signal will certainly not arrive intact.

Fig. 1.2 Rising edge
(black) compared to half a
sinusoid (red). When the
time-domain mirror of the
rising edge is added, the
result looks like one
complete sinusoid

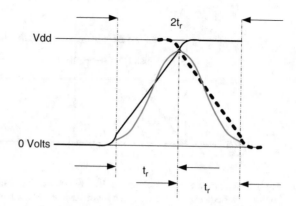

So how can the maximum frequency content be estimated? This is one way to think of it:

The fastest signal in any digital system is the fastest rise or fall time, as discussed above. Now imagine a rising edge followed immediately by a falling edge. The resulting waveform looks like one complete oscillation of a sinusoid. This is illustrated in Fig. 1.2.

The period of the sinusoid is $2t_r$, since the second half is the time-domain "mirror" of the first half, and so the frequency of the sinusoid is $1/2t_r$. We define this frequency as the *knee frequency*, the maximum frequency content of the digital signal. To restate, then,

$$f_{knee} \equiv \frac{1}{2t_r} \qquad (1.4)$$

Not to put too fine a point on it, this is the most important parameter that defines signal integrity. The knee frequency is the highest frequency content of a digital signal.

There is one parameter that is notably absent from the definition of knee frequency – clock frequency! For example, most would assume that the bandlimit of a 3.4 Gigabit per second (Gbps) HDMI connection is 3.4 GHz, because the HDMI signal is transmitting data at 3.4 Gbit/s and the data clock is 3.4 GHz. But this is not correct. The signals have to rise and fall much faster than 1/3.4 GHz (which is 294 ps) to operate correctly – they have to zoom up or down to their final value long before the clock cycle is over. A typical rise time is about 75 ps, and so the knee frequency is $1/(2 \times 75$ ps$) = 1/150$ ps $= 6.7$ GHz. In other words, a typical signal-carrying HDMI data at 3.4 Gbps has a knee frequency of 6.7 GHz.

This is illustrated below in Fig. 1.3.

In other words, knee frequency is of paramount importance, and yet is often misunderstood. The knee frequency of a digital signal is almost always much higher than the clock frequency associated with the signal.

Example 1.4

In a 3 Gbps SATA interface with a rise time of 136 ps, what is its knee frequency?

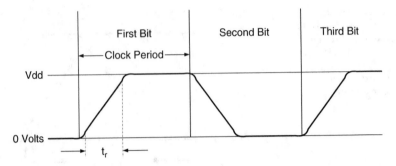

Fig. 1.3 Rise time versus clock period. Rise time is usually much faster than the clock period, and so the knee frequency is usually much higher than the clock frequency

$$f_{knee} = 1/(2*136 \text{ ps}) = 1/272 \text{ ps} = 3.68 \text{ GHz}.$$

3 Gbps SATA has significant signal energy out to 3.7 GHz. SATA cables must work up to that frequency, and you need an oscilloscope with more than 3.7 GHz bandwidth to probe it.

Other textbooks define knee frequency differently. This definition has been chosen because, first, its derivation is easy to remember and, second, it is the highest frequency of all the alternatives. (In other words, it forces you to aim higher.)

Lumped Versus Distributed

The notion of the length of a rising edge leads to another important concept. Way back in introductory circuits class, wires were treated as if they have the same voltage at every point on the wire. A wire could be a mile long and it would be treated that way – same voltage at every point on the mile-long wire. However, Eq. 1.3 shows an expression for the length of a rising edge, and it is commonly found to be on the order of a few inches or a few centimeters. Remember the example – the rising edge was 0.866 cm long, and the trace was 3.4 cm long. Think about what this means: If a rising edge is 0.9 cm long, and the wire is 3.4 cm long, then, at the instant when the rising edge is halfway down the wire, the wire has different voltages along its length. The cozy assumption that the voltage is the same at every point on the wire is no longer true, and any model based on the assumption will be completely wrong. This is illustrated below in Fig. 1.4.

This figure shows what happens as a rising (or falling) edge moves down a longer trace. When the edge is halfway down the wire, the wire has different voltages along its length. Another way to think of it is this – the edge is traveling at a finite speed, below the speed of light. By the time the signal has finished rising at the source, the load (the other end) is still not "aware" that the signal has risen. In fact, it cannot be

Fig. 1.4 Rising edge part of the way down a wire. At the illustrated point in time, the rising edge is partially down the length of the wire

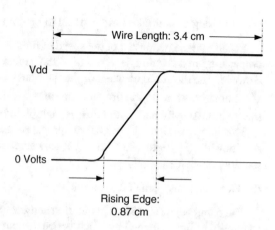

aware – information always travels at or below the speed of light. The speed of light delay causes this situation.

The assumption made back in introductory circuits class was that all circuits are *lumped*. If a circuit is not lumped (e.g., if wires are longer than rising edges), then the circuit is *distributed*. Distributed analysis must include time-domain information about where voltages are located at instants in time, and wires are no longer wires. (They are, in fact, *transmission lines* which we will study in depth in Chap. 8.)

So, what is the boundary between lumped and distributed? There is not a sharp dividing line, but there are some commonly used rules of thumb. The distinction has to do with the length of wires versus the length of the rising edges that run down the wires.

Clearly, if a wire is longer than the length of a rising edge, it is distributed.

But what if a wire is, say, half the length of a rising edge? Or one-third the length? There is still a significant voltage shift along the wire, roughly half or one-third the voltage of a full logic swing. So what is the breakpoint? When is a wire lumped?

One can define the *critical length* as the length of a wire above which the system containing the wire is considered distributed instead of lumped. In this book, we will use the formula that the critical length is $\ell/6$, where ℓ (the length of a rising edge) is as defined above in Eqs. 1.2 and 1.3. So the critical length is one-sixth the length of the shortest (i.e., the fastest) rising edge. The denominator is approximate because the transition between lumped and distributed is fuzzy. Some texts use the denominator 2π. However, 2π sounds very precise, which is misleading, and so this text uses a denominator of 6.

We can work a comprehensive example...

Example 1.5

Consider a 3 Gbps SATA interface, commonly seen connecting a PC motherboard to a hard disk drive. It has a rise time of 136 ps and the signal is routed on a wire that is 15 cm long inside a circuit board made of FR-4 fiberglass, with $\varepsilon_r = 4.2$. (The seemingly random rise time is the maximum allowed by the SATA industry standard.)

(a) How long does it take for the signal to propagate down the wire?

Recall that the reciprocal of the speed of light is 33.4 ps/cm. So if the signal were propagating in a vacuum, it would take 33.4 ps/cm $\times$ 15 cm = 501 ps to propagate. However, the dielectric constant of the fiberglass is 4.2, and so the signal will travel $\sqrt{4.2}$ times slower. So the propagation time is $\sqrt{4.2} \times 501 = 1027$ ps. (Again, it travels more slowly and so the time is multiplied by $\sqrt{4.2}$, not divided.)

This is an example of how the reciprocal of the speed of light comes in handy. Additionally, I like to use "did it get slower or faster?" to remember whether to multiply or divide by $\sqrt{\varepsilon_r}$.

(b) How long is a rising edge on the wire?

The rising edge is 0.136 ns in duration and the speed of propagation is 0.300 m/ns (the speed of light) divided by $\sqrt{\varepsilon_r}$. (Notice the unit conversion from ps to ns, and the use of c in m/ns.) So the rising edge is $0.136 * 0.300/\sqrt{4.2} = 0.0199$ m = 1.99 cm.

(c) Is the wire lumped or distributed?

The wire is 15 cm long and a rising edge is 1.99 cm long. The wire is quite a bit longer than a rising edge, and so must be considered distributed.

(d) How long can a wire be and still be considered lumped?

A rising edge is 1.99 cm long, and so the critical length is 1.99/6 = 0.33 cm. Any wire carrying a 3 Gbps SATA signal that is longer than 0.33 cm is considered lumped and is so small that almost all conductors carrying a SATA signal are distributed.

The last part of the example highlights why we have to use electromagnetics to analyze modern computer systems: The rise and fall times of signals have become so fast that ordinary connections on circuit boards, connections a few inches long, are now transmission lines. (More on that subject later.)

Combining Rise Times

The concept of rise time can show up in several places. We typically think of a rise time as being associated with the output driver of a digital signal, and there is nothing wrong with this association. However, the signal that comes out might pass through a system that, itself, has a different rise time; a very common example is a low-pass filter of some sort. Another example is signals being probed by an oscilloscope – the oscilloscope acts like a low-pass filter and even has its own rise time.

The most common low-pass filter is a resistor-capacitor (RC) filter, a circuit with a series resistor R and a shunt resistor C. This shows up on circuit boards all the time because the circuit-board trace has nonzero resistance and the presence of the trace over a ground plane gives it nonzero capacitance. In other words, every wire on a circuit board is an RC filter.

So what is the rise time of an RC filter?

As you may recall from introductory circuits classes, an RC circuit has a time constant τ equal to R times C. What you may not recall is that the time constant τ is the time for the signal to undergo an e:1 change in voltage. For example, it is the time for a falling edge to drop from 2.718 V down to 1 V.

As we defined it above, the fall time is the amount of time it takes the signal to go from 90% of a logic 1 down to 10% of a logic 1, or a 9:1 change in voltage. So how many intervals of τ does it take to equal one fall time?

When an RC circuit is falling, it is undergoing exponential decay. For a digital signal that starts at voltage ΔV at time $t = 0$ and drops to 0 V exponentially, $v(t)$ is found by

$$v(t) = \Delta V e^{-t/\tau} \tag{1.5}$$

Now consider the same signal passing through voltage $0.9\,\Delta V$ at time t_1 and $0.1\,\Delta V$ at time t_2. We can divide the two points in time by each other, and so we obtain $\frac{v(t_1)}{v(t_2)} = \frac{0.9\,\Delta V}{0.1\,\Delta V} = 9 = \frac{\Delta V e^{-t_1/\tau}}{\Delta V e^{-t_2/\tau}} = e^{-(t_1-t_2)/\tau}$ and, taking the natural log of both sides, $\ln(9) = -\frac{t_1-t_2}{\tau}$. Since $\tau = RC$, the fall time is t_2-t_1, and the natural log of 9 is about 2.2, we have

$$t_f = 2.2\ RC = t_r \tag{1.6}$$

In other words, the rise or fall time of an RC circuit is 2.2 times the resistance R times the capacitance C.

How are rise times combined? If a signal with rise time A passes through a circuit with rise time B, what is the net rise time? Simple addition does not seem right; rising and falling waveforms are not linear. The ideal formula is one where if, for example, A is much longer than B, the combination should be around A.

So if a signal with a rise time of t_{r1} passes through a circuit with a rise time of t_{r2} what is the effective rise time, t_{reff}, when the signal exits the circuit? There is a helpful approximation that is typically used:

$$t_{reff} = \sqrt{t_{r1}^2 + t_{r2}^2} \tag{1.7}$$

Equation 1.7 is a formula that is often used to *combine* rise times. We will refer to the process described in Eq. 1.7 as a "rise time adder," although this is mildly confusing since you are not literally adding the rise times.

Example 1.6

A digital signal has a 100 ps rise time.

(a) What will the effective rise time be if the signal is passed through an RC filter with $R = 10$ ohms and $C = 10$ pF?

The rise time of the RC filter is $2.2 \times 10 \times 10 \, p = 220$ ps. Effective rise time $= \sqrt{220^2 + 100^2} = 241.6$ ps. Stated differently, the 100 ps signal will have a 241.6 ps rise time when it exits the RC filter.

(b) What will the effective rise time be if the signal is passed through an RC filter with $R = 100$ ohms and $C = 10$ pF?

$2.2 \times 10 \times 10 \, p = 220$ ps. Effective rise time $= \sqrt{220^2 + 100^2} = 2202$ ps

(c) What RC constant will yield a signal with a rise time of 200 ps? For $R = 10$ ohms, what C yields the correct rise time?

Effective rise time $= \sqrt{x^2 + 100^2} = 200$ ps. $200^2 - 100^2 = x^2$, so $x = 173$ ps. Since $t_r = 2.2 \, RC$, $RC = t_r/2.2 = 78.7$ ps. For $R = 10$ ohms, $C = 7.87$, or about 8, pF.

(d) Bob used a SATA PHY chip with a $t_r = 50$ ps. On the circuit board, the actual $t_r = 85$ ps. What is the approximate RC of the circuit to which the PHY is connected?

Effective rise time $= \sqrt{x^2 + 50^2} = 85$ ps. $85^2 - 50^2 = x^2$, so $x = 68.7$ ps. Since $t_r = 2.2 \, RC$, $RC = t_r/2.2 = 31.2$ ps.

Parts a and b highlights the property of the rise time adder formula that the final value is close to the larger of the two values if one of the values is much larger than the other. Parts c and d show the value of the formula – it lets one calculate low-pass filters or, given timing information, estimate the parameters of a low-pass filter.

Appendix

1. Find the reciprocal of the following:

   (a) 4 GHz 250 ps
   (b) 250 MHz 4 ns
   (c) 13.3333 ns 75 MHz
   (d) 300 ps 3.33 GHz

2. Express the speed of light. . .

   (a) In mm/ps 0.3
   (b) In ns/in 0.083

3. Consider a signal with a rise time of 225 ps.

  (a) What is the knee frequency of the signal? 1/450 ps = 2.22 GHz
  (b) Consider the signal on a trace over an insulator with $\varepsilon_r = 8$. How long is the rising edge, in inches? 225 ps = 0.225 ns. 30 × 0.225/sqrt (8) = 2.39 cm = 0.94 inches
  (c) How long does a trace have to be in order to be considered "distributed"? 0.94/6 = 0.16 inches

4. The same signal is routed on a circuit board. The output capacitance is 5 pF and the resistance of the gate that drives the signal is 10 ohms.

  (a) What is the rise time of the signal, taking into account the R and C of the output driver? Sqrt(225^2 + $(2.2 \times 5 \times 10)^2$) = 250.4 ps
  (b) What is the new knee frequency, after RC filtering? 1/(2 × 250.4) = 2.00 GHz

5. Consider a circuit board that has an insulator with $\varepsilon_r = 3$.

  (a) If signal A's rising edge is 8 cm long, what is the rise time of signal A? 30 × t_r/sqrt(3) = 8. t_r = 0.46 ns.
  (b) What is the knee frequency of signal A? 1/(2 × 0.46) = 1.08 GHz
  (c) Signal B has a knee frequency of 1.25 GHz. What is its rise time? 1/2 t_r = 1.25 GHz. t_r = 400 ps.
  (d) How long is a rising edge on Signal B? 30 × 0.4/sqrt(3) = 6.93 cm.

6. A circuit board is being designed with $\varepsilon_r = 4.2$. A trace for a slow, infrequently changing signal has a length of 40 cm.

  (a) What rise time is needed for the trace to be considered distributed and not lumped? To be *lumped*, wire length < $\ell/6$. So $\ell > 6$ × wire length. In other words, for the trace to be lumped, a rising edge must be more than 6 times the length of the wire. So the rising edge must be longer that 6 × 40 cm = 240 cm. Using the ℓ calculation, 240 = t_r × 30/sqrt(4.2). Solve for t_r = 16.4 ns
  (b) What knee frequency corresponds to the rise time you calculated in a? 30.4 MHz
  (c) The signal has a rise time of 1 ns. One way to avoid signal integrity issues is to slow the signal down to the point where it is no longer distributed. What RC (i.e., an RC time constant of what value) is needed to slow the signal down to the rise time you calculated in a? sqrt(1^2 + $(2.2RC)^2$) = 16.4 and so 2.2RC = 16.4. RC = 7.44 ns. (Note the units! 1 and 16.4 are in units of ns.)

Chapter 2
Circuit Boards

Background and Objectives

The world of signal integrity often revolves around circuit boards, and many signal-integrity concepts, such as ground bounce and reflections, tie directly into circuit-board structures. This presents a problem – you have to understand circuit boards in order to understand signal integrity. The purpose of this chapter is to understand "circuit board basics" so that you can understand the lingo. This chapter hits the highlights of circuit boards. If you want a deeper insight, Chap. 15 goes into detail about how circuit boards are actually designed.

When this chapter is finished, you should be able to:

- Understand basic packages for components
- Understand the basic parts of circuit boards
- Understand how circuit boards are manufactured

What Is a Circuit Board?

We have all seen circuit boards. It is the board, usually green, that holds all the components that make a device, like a computer or smartphone, work. These boards are always designed by engineers, and, as an electrical or computer engineer, you will need to know how they work to design them yourself.

The board that holds all the components is called a "circuit board" or "printed circuit board" or "PCB" or "printed wiring board" or "PWB." In this book, it is called a "circuit board." It is made of an insulating material, for obvious reasons. The board holds wiring, both in the form of little wires (also called "traces") and in the form of planes or sheets of conductive material. The main purpose of the board is to hold all of the wiring needed to connect all of the components. Rather than doing all the wiring by hand (which was done at one point; it was called "wire wrap"), a circuit

© Springer Nature Switzerland AG 2022
S. H. Russ, *Signal Integrity*, https://doi.org/10.1007/978-3-030-86927-4_2

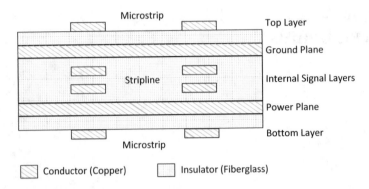

Fig. 2.1 Typical layer stack up

board is fabricated in an optical-etching process that makes all the wires at the same time. It is much more repeatable, much faster, and much more compact than wiring by hand, and results in better (i.e., better-signal-integrity) wiring. Another purpose of the board is structural – it provides mechanical strength and rigidity so that, for example, plugging in a connector does not break the connector.

On a typical modern low-cost circuit board, the insulating material is FR-4 fiberglass and the wires are made of copper. Silver is slightly more conductive but much more expensive, so copper is used instead. Some more expensive boards, requiring higher performance, use other insulating materials.

A typical stack up of signals for a circuit board is shown in Fig. 2.1.

This figure shows a cross-section of a circuit board. Since wires are typically about 1/8th of a millimeter wide, this figure is considerably magnified! There is a lot going on inside a circuit board.

In trying to get all the wires from all of the "point A's" to all of the "point B's," there usually is not enough room on a single layer to connect all the wires. One problem is that wires cannot be too thin – the board would be hard to produce and the electrical resistance would be too high. The wires are designed with some minimum width and spacing, and so only a certain amount of wiring can fit into a certain area. Almost all boards, therefore, have at least two wiring layers. Connecting wiring between different layers are vertical structures called vias. Picturing the board lying flat on a table, a via is a wiring connection that runs vertically from the top of the board to the bottom.

Power and ground can carry a lot of current, and so most modern boards carry power and ground on sheets of conductors called, simply enough, the power plane and the ground plane.

On a typical board, the top and bottom layers are used for signals, the second layer for the ground plane, and the third layer for power. This is a four-layer board, and is found in most consumer electronics; it is a good compromise between cheap and functional. Cheaper boards are of two layers (or even one layer) and more expensive boards are of six or eight layers. (Even though a six- or eight-layer board costs more, it can hold more wiring and therefore leads to a more compact design. That is the trade-off.)

If there are more than four layers, as shown in the figure, the internal layers lie between the power and ground plane, and this leads to an important distinction. Signals on the top and bottom layer have air on one side and the circuit board on the other. This type of routing is called microstrip. Signals in the middle are completely surrounded by circuit-board material and lie between two planes. This type of routing is called stripline. In general, the stripline has more capacitance (the signal is surrounded by circuit-board material, which has more capacitance per unit volume than air) but is almost completely shielded from the outside world.

The layer stack up shown in Fig. 2.1 has four signal layers and two planes, one for ground and one for power. A via is included to show how vias work – they form a vertical copper connection that can connect signals on different layers. In this figure, notice how the ground plane touches the sides of the via, and the power plane does not. In other words, this via is shorted to ground. This structure is often used if a grounded signal is needed on another plane, such as bringing a ground to a component on the top side. The layer stack up in this example is a six-layer board, with the top layer considered layer 1, each plane counting as a layer, and the bottom layer considered layer 6. A four-layer board is similar, but without the two internal signal layers. There is almost always an even number of layers, for reasons discussed below.

Some of the dimensions of the circuit board are constrained by the company that fabricates the board. For example, the signal traces must have some minimum width and some minimum spacing. A typical limit is 5-mil-wide traces with 5-mil spacing. 1 mil is 0.001 inches (1/1000th of an inch) or 0.0254 mm. A 5-mil trace is 0.127 mm wide. The thickness of the signal layers and planes (i.e., the copper thickness) is specified by the designer, in archaic units of "ounces" (more on which below). The thickness of the insulator is specified so that the signal lines have a known characteristic impedance (which we will study layer). Specifying the insulator and conductor thickness is called a layer stack. Unfortunately, many of the units in the circuit-board world are in Imperial units (inches, ounces, mils, etc.) and so this book uses those units.

How Are Circuit Boards Made?

The process starts with a sheet of fiberglass coated on both sides with copper. The thickness of copper is specified by the designer of the board, and there are a few standard thicknesses that are commonly used. Perhaps the most common thickness (and the one that will be used if you do not specify anything) is "1-ounce copper" which is 1 ounce (weight) of copper per square foot of board. In SI units, it is a copper layer that is 35 μm thick. For digital signals, the thickness is surprisingly irrelevant due to the skin effect (more on which later).

The copper is covered in goo and exposed to light. The goo is a material called photoresist that becomes soluble in acid when exposed to light. So if you place a mask in front of the light, the light-dark patterns on the board cause some of the photoresist to become soluble. In other words, light shines through the mask and creates a shadow on the top of the board. The goo becomes soluble where there is light, and remains insoluble where there is no light. Specifically, the mask is dark in the same pattern as the wiring on the layer. So the photoresist becomes soluble wherever it is exposed to light and not in places that are in the shadows.

The process is repeated on the other side of the sheet. (Recall that both sides are covered in copper.)

The structure with exposed photoresist is then dipped in acid. The acid dissolves where the photoresist was exposed to light, and then keeps going and dissolves the copper underneath. Where the photoresist was not exposed, the acid does not etch through and the copper remains.

The result is a two-layer board with wiring patterned on both sides. Each wire on a circuit board is called a trace. The only difference between a plane, like a ground plane, and a signal layer, full of traces, is the amount of copper that is removed.

To make a four-layer board, the company fabricates two two-layer boards and then presses them together. Specifically, a piece of fiberglass impregnated with glue (called pre-preg) is placed between the two two-layer boards, and the stack is put under pressure and heated (or laminated). Think of it as a heated sandwich with pre-preg as the meat and exposed, etched two-layer boards as the bread.

If the board has more than four layers, the process is the same except that the stack is higher and there are more pre-preg layers. It should be clear now that almost all boards have an even number of layers.

What happens next is a little quaint. The vertical structures (specifically the vias) are made by drilling holes (one hole per via). In other words, every individual via is drilled out. Another round of photoresist is applied and etched, and this time copper is added. Copper is added where there is no photoresist (more accurately, removing the photoresist removes the unwanted copper) and so all of the drilled holes become filled with copper. If vias are small enough, the copper completely fills the hole. Larger vias are actually tube-shaped.

Vias are important to understand. If the ground plane touches the part that gets drilled, the via will be shorted to the ground plane. Sometimes this is desirable; for example, this is how you get a ground connection up to a circuit component like a resistor. If the connection to ground (and/or power) is undesirable, then there has to be a hole in the ground and power planes so that, even after the drill bit makes a new hole, the power and ground plane does not touch the resulting via. Again, this is illustrated in Fig. 2.1.

Sometimes the copper-filled hole is placed so that the lead of a circuit component can fit through it. In this case, the via is drilled wider so that, even after coating it with copper, a pin can still fit through it. This structure is called a plated through-hole. A plated through-hole is a fat via, and can be used to route signals from top layer to bottom, connect to power or ground plane, etc.

The result of all of this is a complete, three-dimensional wiring structure with wires and power and ground planes, ready to hold components. Some of the components lie flat on top of the board, and are called surface-mount components. Others have pins that stick through the plated through-holes (and are soldered on bottom) and are called through-hole components. The place where the component is located is called the component's footprint.

Three more layers are typically also added.

First, the structure is covered in a coating (usually green) called solder mask. (The green color we associate with circuit boards is actually the color of the solder mask, not the color of the fiberglass; FR-4 fiberglass is a tan color.) This coating keeps the solder from sticking to parts of the board where solder is undesirable. Stated differently, it only leaves copper exposed in places where solder is desired, like places where solder touches circuit components. The solder mask has some signal-integrity consequences. It adds a layer of insulator to the top of most conductors and increases the dielectric constant (i.e., capacitance) in subtle ways.

Second, a layer of markings is added in order to make the board human-readable. This is applied to the board in a silk-screen process (yes, the same process used to put designs on T-shirts) and is called silkscreen or sometimes just silk. Typical silkscreen markings include labels to identify components (called reference designators or ref des's), markings to indicate part orientation (e.g., diodes), and sometimes the name of the company or warning labels. These markings are essential; it is the only way the factory knows which part is which and which orientation is correct. It is also the only way a repair technician knows where to look, and the only way a design engineer can check the design. Warning labels are important, too – they can flag places that are dangerous to touch. Let's just say that I learned this the hard way, and asked the engineers in my team to add warning labels.

Third, solder is sometimes placed where components will later be soldered. When a board has solder added, it is said to be tinned. You can recognize the solder because it looks like (and is) shiny metal. A complete bare circuit board, ready to have parts added to it, is shown in Fig. 2.2.

There is a lot to see in this picture.

The trace looks like it is light green on top of dark green. In reality, a copper trace lies over the olive-colored fiberglass, and the green solder mask gives everything the green tint. This board has not been tinned; tinning is normally silvery colored and is thick. The copper rectangles would look like small silver pillows if the board were tinned.

The plated through-holes are present so that the pins of through-hole parts can stick through them. They are easy to spot because they are much larger than vias. The vias are much smaller and can be hard to see. They are just big enough to make a track running from top to bottom that can be filled with copper.

The pads are rectangles of copper ready to have surface-mount components attached to them. The pad is the place where a surface-mount component's signal lead makes contact with the board's wiring. If you look closely, you can see traces exiting most pads. When the board is tinned, the pad becomes a pillow-shaped rectangle of solder.

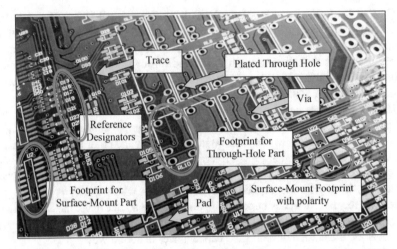

Fig. 2.2 An example of a bare circuit board. The green color is the solder mask and the white labeling is silkscreen. Note the board holds both surface-mount and through-hole parts

The reference designators are the white labels that are silk-screened onto the top. Resistors start with R, diodes with D, etc. (There is an informal convention for reference designators.) The designer of the board also added features like white rectangles to show where parts go.

If you look at the rectangles and see where components go, you can see the "footprints" where parts go. The surface-mount footprints are arrays of pads. Surface-mount parts lie on top of the pads and are soldered down. The silk screen of the small surface-mount footprint has a polarity marking – the small notch at one end of the rectangle. This makes clear which way the part needs to face before it is soldered down.

Components and Component Packages

Circuit components come in a variety of package styles. Components include not only passive components, like resistors, capacitors, inductors, and diodes, but also integrated circuits and transistors. Passive components typically have two terminals or wires, and integrated circuits can have dozens or even hundreds of connections.

Sometimes the component has little legs (or stiff wires) that stick through the board. The component's legs sit in plated through-holes (Fig. 2.2) and are then soldered from below. This type of part is called a through-hole component (Fig. 2.3).

The pins or wires add inductance to the signals, and often parts have to be placed by hand, because machines can easily bend the pins.

Sometimes the part is basically flat and just sits on the top of the board. This is called a surface-mount part (Fig. 2.4).

Fig. 2.3 Through-hole components

Fig. 2.4 Surface-mount components

These parts are small and place on a circuit board by machine. The smaller packages are less expensive, and the smaller size leads to lower inductance.

ENGINEER'S NOTEBOOK – MORE ABOUT COMPONENT PACKAGING

The simplest packages are the leaded packages used for resistors and capacitors. In the case of resistors, these are the brown resistors with colored bands that most students use in the first electrical engineering lab. They are called "leaded" because they have leads. (Note that this is the term "lead" with a long e – meaning a wire that sticks out – and not lead with a short e – meaning the toxic metal.) Capacitors can have leads sticking out both ends (axial) and both leads sticking out of one end (radial).

Chips that are leaded come in packages with pins that stick out of the bottom. The simplest is the dual-inline package or DIP, which is the package that is used in most introductory digital labs. There are other through-hole chip packages, but they are increasingly rare.

Most connectors come in through-hole packages, so that the mechanical strain of inserting and removing the mating connector is transferred to the circuit board.

Surface-mount packages, as mentioned above, are designed to sit on top of a circuit board.

The simplest are the two-terminal packages for resistors and capacitors. They look like little shoeboxes with metal on both ends. The package sizes are called out in hundredths of an inch, and so a 1206 package is 0.12 inches by 0.06 inches. Common sizes today are 0805, 0402, and 0201. (0402 is considered the smallest that can be soldered by hand, and even then only with a very steady hand.) 01005 is on the horizon, which is 0.01 inches by 0.005 inches.

Chips come in packages with leads that stick down and then out. The result is a set of pads that lie horizontally, with the number of pads equal to the number of circuit connections that are needed. They range from 3-pin packages, such as an SOT-23, up to hundreds of pins, so-called quad flat packs or QFPs.

In newer versions, there are no leads that stick out; rather the package is arranged so that there is a tiny square of metal next to each integrated circuit pad. These are sometimes called chip-scale packages and are very, very small. (Most cannot be soldered by hand.)

Chips with more connections come in packages with an entire array of "dots" on the bottom. The dots look like little balls and so the package is called a ball-grid array or BGA.

Smaller packages transmit heat better, allow more circuits to be crammed into a tinier space, and are more difficult to manufacture and rework. (Some rework is always necessary because the assembly machines are not perfect.) There is no clear correlation of package size to price – it really depends on the manufacturer's volume – but it is generally the case that more pins or balls, more the cost.

How Are Circuit Boards Used to Make a Product?

The circuit board is the focal point of almost every electronic design because it contains all of the wirings and holds all of the components. So what happens next is that a horde of skilled laborers has to put all the right components in all the right orientations in all the right places onto every single board. Or else your company will go broke.

The way that it is done is with machines called pick-and-place machines. These machines are loaded with reels of components. If you designed a board with a 1N914 silicon diode, a reel of 1N914 diodes is loaded onto a location on a machine. So the process of assembling a board starts earlier than one might think – it starts when the technicians load the reels onto the machine. This is a tedious and error-prone process, and it leads to the unfortunate consequence that if one part is wrong, it is usually wrong on every board.

Every time a new board enters the machine, a vacuum head reaches over and sucks up a single diode (in this example), pulls it off the reel, goes over to exactly the right place on the board, rotates the diode into the correct orientation, and then "poof" shoots it onto the board. In a cable set-top box, for example, this process is

Fig. 2.5 Successful surface-mount process. Each of the components pictured here is 0805, meaning 0.08 inches by 0.05 inches (2 mm by 1.3 mm), and has eight surface-mount leads on the bottom

repeated for about 2,000 components. Fortunately, this process is a lot faster than it sounds. Also, the machine usually has several vacuum heads.

What makes the process amazing is when you stop to think what can go wrong. The part can land in the wrong place. It can be in the wrong orientation. It can be the wrong part (the wrong reel or the wrong slot). Even if everything is correct, the machine can make a mistake and the part is not lined up correctly when it lands, for example, if the machine does not shoot it straight. Making life more difficult, these parts are often incredibly small. For example, a typical component size is "0402" meaning that it is .04 inches by .02 inches (1 mm by 0.5 mm) in size. Modern boards can have 0201 or even 01005 components on them.

Once all the parts have been placed, the entire board passes through a reflow oven which looks a lot like a pizza oven. The board rolls through on a conveyor and different zones in the oven create different temperatures. The goal is to slowly heat the board to just below the point where solder melts, rapidly take it above and below the melting temperature, and then slowly cool it back to human-touchable temperature. This process melts the solder and simultaneously solders all of the components onto the board at once. (Thought question: So how do you put surface-mount components on both sides of a board?) (Fig. 2.5)

The through-hole assembly process is more straightforward. The parts are inserted into the holes. Small parts, like capacitors, are inserted by machine. Complicated parts, like high-definition television connectors, are often inserted by hand. The leads are trimmed so that they only stick out a short distance below the board. Then the board is passed over a wave solder machine. The wave solder machine contains a lake of molten solder. (In fact, the solder looks solid until you whack the side of the machine with your hand, and it undulates. Think Terminator 2.) The solder flows over a ridge that rises out of the lake of solder, and each board barely touches the flowing solder. The flowing solder is called a wave, hence the name wave solder. Every lead that sticks out of the board sucks the solder up through capillary action and all the leads are soldered (Fig. 2.6).

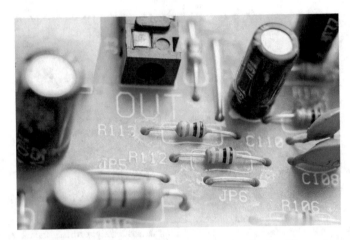

Fig. 2.6 Through-hole board. This particular board is so inexpensive that it has no traces on the top side. The bare wire JP6 is being used to jump over a trace on the bottom side

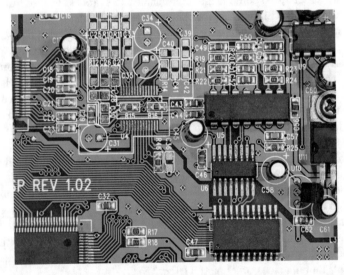

Fig. 2.7 Board containing both surface-mount and through-hole parts. Note how Capacitor C31 and chip U4 have no components. This is called a population option – this particular version of the board did not need C31 or U4 and it was built without them to save money

So what if you need to add a through-hole part to a surface-mount board? The solder wave would suck all of the bottom-side surface-mount parts back off the board, with tragic consequences. So a board is laid inside of a selective-solder pallet which only has holes where there are through-hole parts. This leads to an important but obscure design rule – through-hole components must be kept a minimum distance from bottom-side surface-mount parts so that there is room for the selective-solder-pallet hole (Fig. 2.7).

It is very important to understand how boards are assembled in order to understand how to lay them out correctly and how to test them. For example, the selective-solder pallet creates spacing rules between bottom-side surface-mount parts and through-hole parts. Surface-mount parts can easily be put on backwards, and parts with the same footprint can get mixed up.

Once the board has all the components on it, it is ready to test.

Chapters 15 and 17 discuss the design and debug process in more detail.

Appendix

1. Consult online sources and find the layer stack for a four-layer circuit board (copper layers on top and bottom, power and ground planes in the middle) that is 63 mils thick. (This is a standard four-layer circuit board.)
2. What does 1-ounce copper mean?
3. A co-worker shows you a resistor that is a very small rectangular shape. Is it through hole or surface mount?
4. What are the steps in manufacturing a circuit board?
5. Define the following circuit-board terms:

   (a) Trace
   (b) Plane
   (c) Via
   (d) Through-hole
   (e) Surface-mount
   (f) Solder mask
   (g) Selective-solder pallet

6. How are through-hole parts usually soldered onto a circuit board?
7. How are surface-mount parts usually soldered onto a circuit board?
8. Why does a board with 50 vias cost more than a board with 20 vias?
9. What keeps a via from shorting to a plane?
10. Bob is designing a six-layer board and has used a via to route signal A from the top side to the bottom side. Why can't any other signal be routed on an internal signal layer at the location of the via?
11. What features does a designer add to the silkscreen to help indicate how parts are oriented?
12. How big is a 01005 component, in inches and in mm?

Chapter 3
Gates, Packaging, and Boards: Properties and Modeling

Background and Objectives

This chapter connects packaging and circuit boards, which were discussed in the previous chapter, to logic gates (actually, logic inputs and outputs) and reviews concepts like noise margin, drive strength, and source resistance. This chapter also discusses heat and temperature and discusses the different options that exist for modeling boards and circuits. When you are finished with this chapter, you should be able to:

- Discuss the effects of circuit structures on signals
- Calculate temperature rise based on packaging parameters
- Describe ways of modeling circuits and decide which method is more appropriate
- Construct simple circuit models in simulation tools such as SPICE

What Limits What Is Possible?

We have all seen electronic products steadily shrink in size and increase in complexity. It is not too difficult to imagine wearing an entire smartphone in our ear. So the question is, what limits the size and speed of digital circuits? What keeps them from being smaller and faster? There turns out to be four factors that impose limits.

The first limit is that transistors are of a nonzero size. In fact, their size is limited by the current fabrication technology. As of the writing of this book, the limit of fabrication technology is at a node name "7 nm." At about 5 nm, conventional circuit fabrication is expected to reach its fundamental limit. Keep in mind that at 5 nm, the gate of a transistor will be about 25 silicon atoms in length. It is difficult to imagine a gate that is much shorter than that! What is worse, the gate length is reaching the point where the probability function that represents the location of an electron extends across the gate. In this spooky quantum realm, electron tunneling will

© Springer Nature Switzerland AG 2022
S. H. Russ, *Signal Integrity*, https://doi.org/10.1007/978-3-030-86927-4_3

make it impossible to turn gates completely off. In other words, we are rapidly reaching a point where the size of a transistor is no longer constrained by the limits of fabrication; it will be constrained by fundamental physical limits.

The second limit is that transistors have a finite drive strength, which is, in turn, a function of the fact that their internal structure has capacitance and resistance. They can only source or sink a finite amount of current, and this limits how fast signals can change. As one might expect, smaller transistors typically have lower drive strength, and so making the transistors smaller also makes them weaker. The result of this limitation is that logic gates respond in a nonzero amount of time and, when they source or sink current, do so with nonzero electrical resistance.

The third limit is the interconnection between transistors. The first interconnection limit is the speed of light. Signals must travel at finite velocities, constrained by the dielectric material (insulator) surrounding the wire. As we have seen, the speed of propagation is roughly the speed of light divided by the square root of the relative dielectric constant of the dielectric material. This is how I remember it: The speed of light is 1 foot per nanosecond. A ball 1 foot in diameter (say a volleyball) is the volume of space that can be accessed (send out a signal and receive a response) in one nanosecond. Factor in the dielectric constant, and the ball shrinks. The second interconnection limit is that practical interconnections also have capacitance and resistance. The capacitance is unavoidable and troublesome. It is unavoidable because a signal wire over a ground plane forms an intrinsic capacitor. It is troublesome because the resistance of the logic gate (plus the resistance of the wire) times the capacitance of the signal wire forms an RC circuit with a time constant. It does not matter how fast the gate switches; the signal swing will always be limited by the RC constant.

The fourth limit may be surprising, but it is heat. All of this process of changing signals from high to low requires real energy, and the more tightly one packs in circuits that change, the higher the heat density. There are thermodynamic arguments that come into play here. To make a long story short, the processing of information necessitates the dissipation of energy, and the dissipation of energy must, of physical-law necessity, create waste heat.

Heat is already more of a factor than most people realize. The datasheet for an Intel Xeon processor shows a maximum power dissipation of 155 Watts. This is 155 Watts being dissipated in a piece of silicon roughly the size of your thumbnail. This is the same amount of heat as being produced by a very bright incandescent floodlight, like the kind you use to keep a tray of food hot, except that you are trying to keep the Xeon cool. Little surprise, then, that liquid-cooled computers are coming back in vogue. (I literally met a man in Walmart who is keeping his computer cool using mineral oil and a radiator from an old Mazda.)

In fact, it is a physical law: If a computer is not constrained by heat then it has not reached its fundamental minimum size. This is so because heat winds up being the practical limit on how tightly circuits can be packed together; it is the one constraint on size that is unavoidable. (Consult thermodynamics texts on why this is so.)

So if there was a perfect logic family that had gates with zero resistance and capacitance would it still dissipate power? The answer is yes. Consider Eq. 3.1 for the current in a capacitor:

$$I(t) = C dV(t)/dt \tag{3.1}$$

Consider now a signal wire with capacitance C making a logic swing from 0 Volts to Vdd Volts. (Vdd is a common abbreviation for the power voltage of a digital system.) Recalling that power is current times voltage and that energy is the integral of power over time, we have

$$E = I(t)V(t) = CV(t) dV(t)/dt \tag{3.2}$$

$$E = \int_0^{Vdd} CV(t) dV/dt \, dt = \frac{1}{2}CV^2 \Big]_{V=0}^{V=Vdd} = \frac{1}{2}CVdd^2 - 0 = \frac{1}{2}CVdd^2 \tag{3.3}$$

The energy required to charge a capacitor from 0 to Vdd is $\frac{1}{2}CVdd^2$. This energy requirement is intrinsic to the capacitor; even a logic gate with 0 internal resistance and capacitance dissipates this amount of energy. (This leads to an important point – this energy is being dissipated in the gate driving the capacitor, not the capacitor itself.) If a gate toggles with frequency f, then the power dissipation P is found by

$$P = Ef = fCVdd^2 \tag{3.4}$$

The frequency f in the above equation is quite misleading. It is the frequency at which the particular signal *toggles*. It is specifically *not* the clock frequency. In a typical integrated circuit, like a microprocessor, a single signal toggles relatively infrequently, and so f can be a small fraction of the overall clock speed.

There are two important lessons from this equation, however. The first is that power dissipation is inevitable because interconnections are capacitive. The second is that power dissipation increases linearly with frequency and at the square of the supply voltage.

There are other sources of power dissipation, such as the aforementioned internal resistance and capacitance of logic gates, and the nonzero resistance of the capacitor being charged and discharged. Each transistor may also "leak" current when turned off. This is outside the scope of the book, but suffice to say that power dissipation only goes up from Eq. 3.4.

Power and Heat Dissipation

All of the toggling dissipates power, and so it ultimately winds up as heat. What happens to all of the heat?

We can conduct a thought experiment. Let's pretend that we take a resistor and put it inside a small box, along with a temperature sensor. We can set the resistor to dissipate 1 Watt of power and it will get hot. (It will get surprisingly hot. 1 Watt turns out to be a lot if you touch it.) Say, for the sake of discussion, that the ambient temperature in the room is 20 °C and the resistor reaches a temperature of 50 °C.

What happens if the ambient temperature goes up? For example, what if the room heats up to 25 °C? What happens is that the resistor goes up by the same amount; it is simple physics. So the resistor heats up to 55 °C.

What happens if the power dissipation goes up? It turns out that the temperature difference between the temperature of the resistor and the temperature of the ambient air is a linear function of power. The constant of linearity depends on the size and shape of the resistor, the packaging around it, and whether or not the air is moving. So if the power dissipation went up to 1.5 Watts, with an ambient of 20 °C, the temperature of the resistor goes up to 65 °C. In other words, the temperature of the resistor *above ambient* goes from 30 to 45 °C.

We model this phenomenon by defining a constant, the *thermal coefficient* or θ, which has units of °C/W. In our example, the resistor rose to a temperature of 30 °C above ambient with a power dissipation of 1 Watt, and so θ = 30 °C/W.

To recap, the final temperature of a heat-dissipating element is some offset from ambient. If the ambient temperature goes up 10 °C, the temperature of the element goes up 10 °C. The offset is a linear function of the power being dissipated, modeled with a thermal coefficient θ.

$$T_{\text{element}} = T_{\text{ambient}} + \Theta P_{\text{diss}} \qquad (3.5)$$

where T_{element} is the temperature of the heat-dissipating element, T_{ambient} is the ambient temperature, and P_{diss} is the power being dissipated.

The "element" is important to understand. For example, if there is a logic gate inside a package that is toggling, the "element" is composed of the transistors that make up the logic gate. The transistors themselves get hot. The good news is that silicon transistors can get hot, on the order of 125–150 °C. The bad news is that there is a lot that can trap the heat.

For most systems, the total temperature rise is a function of two effects. The first is that the package surrounding the electrical component adds heat (i.e., it has its own thermal coefficient). When you look at a "chip," the black rectangle you see is actually the package around the actual integrated circuit. This package traps heat. The second has to do with whether the package has heat-dissipating or heat-spreading features such as a heatsink or moving air.

Since another name for a transistor is "junction," the thermal coefficient can be broken up into two components, the junction-to-case thermal coefficient (θ_{JC}) and the case-to-air thermal coefficient (θ_{CA}). The total, overall thermal coefficient, then, is θ_{JA}. That is,

$$\Theta_{JA} = \Theta_{JC} + \Theta_{CA} \tag{3.6}$$

Armed with this knowledge, we can work on the below examples.

Example 3.1

Consider a microprocessor with a thermal coefficient $\theta_{JA} = 25\ °C/W$ and an absolute junction temperature limit of $150\ °C$. ($150\ °C$ is a typical temperature limit.)

(a) With the processor dissipating 1.5 W of power and at an ambient of 25 °C, how hot do the transistors in the microprocessor get?

$$T_{junction} = T_{ambient} + P_{diss}\theta_{JA} = 25 + 1.5 \times 25 = 62.5\ °C$$

(b) At an ambient of 125 °C (common in the automotive industry), how much power can it dissipate?

$$T_{junction} = T_{ambient} + P_{diss}\theta_{JA}.125 = 100 + P_{diss} \times 25, \text{so } P_{diss} = 1.$$

(c) If it was desired to operate the chip with 1.5 W of power dissipation at 125 °C ambient, what thermal coefficient is required?

$$150 = 125 + 1.5 \times \theta_{JA}, \text{and so } \theta_{JA} = (150 - 125)/1.5 = 16.6.$$

So how do you lower the thermal coefficient?

In terms of the case (θ_{JC}), it involves shrinking it (making it as close to the size of the actual die containing the transistors as possible) and fabricating it from materials that are more thermally conductive (in other words, more expensive). Ceramic packages are used instead of plastic when this is of concern.

In terms of the ambient (θ_{CA}), it involves adding a heatsink and possibly forced air. A heatsink is straightforward, although it must be attached with thermally conductive material (so-called heatsink glue) and must be fastened down if there is concern above vibration or handling. Forced air (typically adding a fan) is a major design change. For example, in the automotive example above, it probably is not practical, nor would it be practical in a handheld device such as a cell phone. Conversely, in some settings, such as servers, forced air is considered routine.

Summarizing the Effects

Chip packages and circuit-board interconnections can cause signals to undergo degradation. This can occur due to the properties of the wiring and the packaging. The effects of the wiring are easily understood – physical wires have resistance, inductance, and capacitance. The packaging has more subtle effects. For example, the package forces the power and ground connections to be routed through a small number of package pins, increasing the inductance and resistance. The pins can be close together, increasing the capacitance.

One way to take all of this into account is through modeling and simulation.

Simplest Gate Model

Consider a complementary metal-oxide-semiconductor (CMOS) logic gate output driving a CMOS logic gate input. (We are considering CMOS because almost all digital logic today is constructed from it.) The output of a CMOS logic gate has one or more N-channel and P-channel *field effect transistors* (FETs). The N-channel FET is capable of shorting the output to ground and the P-channel FET is capable of shorting the output to *Vdd*. (*Vdd* is the power supply voltage of the gate.) The short is not perfect – as noted above all transistors have some resistance – but in the case of CMOS, the resistance can be quite small (milliohms to ohms).

This leads to a very simple, and surprisingly accurate, model of a CMOS output. Each transistor is modeled as a switch in series with a resistor. The switch represents the on–off behavior of the transistor (to be specific, the ability of the channel to conduct or not), and the resistor represents the "on" resistance of the transistor (a combination of the electrical resistance and its ability to source or sink current).

The input of a CMOS logic gate is a signal that is routed to the gate of an N-channel and P-channel transistor. CMOS gets its name from the fact that the gate is a sandwich: A layer of Metal over a layer of (insulating) Oxide over a layer of Semiconductor. To make a long story short, it looks electrically like a capacitor. Since the logic-input stage contains wiring to both *Vdd* and ground, it looks like a capacitor to *Vdd* and a capacitor to ground. The result is shown in Fig. 3.1.

Even this simple model demonstrates some important principles. For example, consider what happens when the output switches from high to low. When the output was high, it was shorted to *Vdd*. This discharged the capacitor on the top to zero volts and charged the capacitor on the bottom to *Vdd* volts. When the output switches, the P-channel FET (the one on top) open circuits and the N-channel FET (on bottom) becomes a short circuit. The capacitor on the bottom discharges through the switch and the resistor.

The current I associated with the discharge (i.e., the actual flow of charge), flows in a loop from the capacitor, through the signal wire, through the switch and resistor, and then *back through the ground connection* to the other side of the capacitor. The

Fig. 3.1 Simple model of a
CMOS output driving a
CMOS input

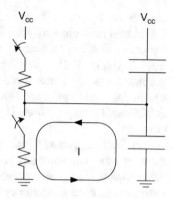

current flows in a loop, and the loop is completed by the ground of the power supply. Likewise, when the signal switches from low to high, the loop is completed by the *Vdd* of the power supply. This loop-completion portion carries the *return current* of the signal.

This property of switching logic gates – that the circuit is completed through the power supply – is one of the most misunderstood elements of digital design. The signal integrity of the ground connection is every bit as important as the signal integrity of the signal wiring, because the ground connection carries the return current.

This also leads to a frequently misunderstood concept, that of return current. Every signal needs a path for the current to return. Usually, this path is on the ground plane. The return current is real (i.e., it actually exists) and needs to be taken into account. Anything that disrupts the path of return current (e.g., detours it and makes it longer) or any sensitive circuit near the return current is a possible signal-integrity issue. It is very easy to remember the wire – it is necessary to carry the signal – but remembering the return path is equally important.

Electrical Modeling

In simulating modern integrated circuits, which can hold millions of logic gates, a simplified model is used in order to make simulation computationally feasible. Each logic gate is modeled as a time-delay element, with the minimum and maximum propagation delay noted for each gate. This is similar to the switch-and-resistor model, except that time delays are inserted instead of resistance. Even with this simplification, the simulation of a chip can take days.

There are problems with these models. For example, transistor operation is much more complicated than a wire and a resistor, and a logic input looks more complicated than a couple of capacitors.

One common method for increasing the accuracy of a simulation is to use a more accurate circuit simulator, like SPICE or one of its variants. SPICE simulation is considered an extremely accurate model, and is used when designing complex integrated circuits to verify clock-cycle times (i.e., maximum operating frequency). However, SPICE is so much more complicated than a simple delay model that it is used sparingly. Each type of gate is typically SPICE-simulated to determine its minimum and maximum speed, and then SPICE is only used to simulate the few circuit paths that limit clock speed (the slowest paths). The important point is that SPICE can be very accurate if you can calculate the circuit-simulation parameters correctly.

This leads to the main problem with SPICE, determining the circuit parameters. Some of them, like those that govern transistor behavior, can be determined from the fabrication process (and detailed knowledge of how transistors work). Others can be extracted from the geometry of the wiring (e.g., by using the length and width of a wire and its height above a ground plane, its capacitance can be calculated).

There is a subtle problem with SPICE if all the parameters are known. If a company has its own fabrication process, it will not want to disclose the SPICE model – the model contains too much proprietary information. Some in the industry have switched to IBIS models, which are simpler models that do not disclose proprietary data. However, not all chips have IBIS models, especially those that are custom-made.

Calculating parameters like capacitance can also be quite complicated. For example, on a circuit board, the capacitance is a function not only of the geometry of the signal wire but also of the composition (and thickness) of the dielectric and of the solder mask.

One way to calculate these parameters more accurately is by using a field solver. A 2-D field solver can solve the cross-section of a circuit and works for structures that are relatively long (have a relatively constant cross-section). If a structure twists and turns or has elements like vias, a 3-D field solver may be needed.

This all adds up to a hierarchy of modeling.

When starting a complex, high-speed layout, one might simply use the design guidelines provided by the manufacturers of the ICs. In fact, there is a lot of value in doing this, not the least of which is that the manufacturer is responsible (to some extent) if it doesn't work.

Whether or not modeling is useful and warranted depends on a series of issues. SPICE is relatively easy to set up (once you get the hang of it), but are all the SPICE parameters known? If not, how hard is it to calculate the parameters? For example, is a field solver needed to calculate the SPICE parameters accurately? If one does not think all this through ahead of time, one can plunge into a black hole of modeling and simulation.

The important point is to determine upfront how much is known and how much needs to be simulated. Once the size of the simulation work is scoped out, an informed decision can be made as to whether the modeling work is worth the effort.

If one is working with a new technology (like a new generation of memory technology, or a new bus standard), or if one is working in a marketplace with a low volume of relatively expensive, performance-intensive products (e.g., the military market), then a lot of modeling and simulation might be cost-justified. In consumer electronics, with emphasis on low-cost products and rapid time to market, extensive modeling and simulation may not be warranted.

The Modeling Process

So how does modeling and simulation typically play out?

In terms of designing an IC, extensive modeling is used to perform timing analysis and extensive simulation is used to confirm the circuits work correctly. This is largely outside the scope of this textbook.

In terms of designing a complex multi-chip electronic product, there is a three-step process.

First, one should apply knowledge of physical laws in order to "do it right the first time." For an experienced designer, this is more straightforward than it sounds – the overwhelming majority of circuit interconnections are not complicated. And when you are finished with this book, you will have a grasp of the physical laws.

Second, schedule and manage projects with knowledge of likely mistakes. We will discuss project management in more detail later, but suffice to say that the skill lies in identifying the parts that will be hard and starting with those elements first. Knowledge of likely mistakes, like schematic or layout errors, will enable everyone to double-check designs skillfully.

Third, if the analysis (or subsequent design efforts) identify signals and parts of circuits that are troublesome, marginal, challenging, or extremely important, it may be worth the effort to model and simulate those signals. High-speed interconnects, power infrastructure, and places in the layout where the optimal layout will not fit (forcing one to use a sub-optimal layout) are examples.

So how are models constructed?

One method is to measure a sample. For example, one could do this with a capacitor. If you connect a capacitor to a variable-frequency power supply and measure the absolute value of the impedance, it will drop as the frequency goes up. But then the unexpected happens. At some point, the impedance will reach a minimum and then it will start going back up. This is illustrated below in Fig. 3.2.

Fig. 3.2 Impedance of a capacitor versus frequency

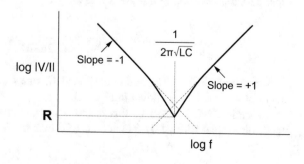

Fig. 3.3 Model of a
capacitor

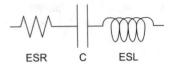

ESR   C   ESL

The vertical axis is the log of the magnitude of the voltage divided by the current, also known as the magnitude of the impedance. The horizontal axis is the log of frequency.

What is going on? It turns out that a capacitor is not a capacitor. It has capacitance – that is, the physical origin of the straight line on the left that goes down as frequency goes up. (On a log-log scale, the graph of impedance versus frequency has a slope of -1.) But the capacitor also has resistance; that is, the minimum impedance value of the curve (shown by the letter R). The capacitor also has inductance, and that is the physical origin of the straight line on the right that goes up. The resistance is called the *equivalent series resistance* or ESR and the inductance is called the *equivalent series inductance* or ESL. If you go back through the math, the frequency of minimum impedance is the point where the impedance due to capacitance and due to inductance exactly cancel. Specifically, since

$$Z = R - \frac{1}{j\omega C} + j\omega L \tag{3.7}$$

Z is minimized when

$$\frac{1}{j\omega C} = j\omega L \tag{3.8}$$

or

$$\omega = \frac{1}{\sqrt{LC}} \tag{3.9}$$

This point is called the *self-resonant frequency* or SRF. Since many datasheets only give the ESR of a capacitor and the SRF, one can calculate the ESL from the other parameters. Solving Eq. 3.9 for L, we get

$$ESL = \frac{1}{C\omega^2} = \frac{1}{C(2\pi SRF)^2} \tag{3.10}$$

Armed with ESL, ESR, and capacitance C, one can then construct a detailed model of a capacitor, as shown in Fig. 3.3.

A simple SPICE model of this circuit can be constructed. Consider a 10 μF capacitor with an ESR of 0.05 Ω and an ESL of 2 nH (Fig. 3.4).

Fig. 3.4 SPICE model of a
capacitor

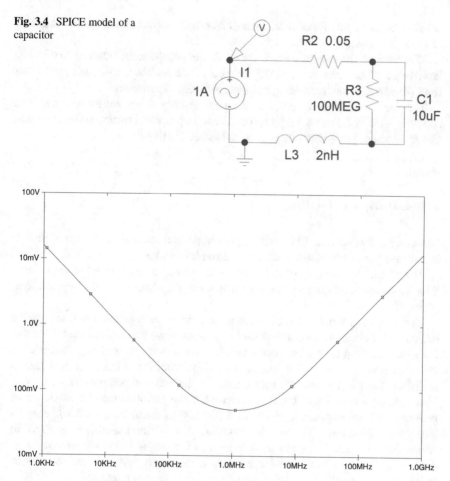

Fig. 3.5 SPICE simulation results

The model uses an AC current source set to a value of 1 Amp. This current source is being used to measure impedance. Since $V = IR$, if $I = 1$, then $V = R$. In other words, the 1 Amp current source creates a situation where the voltage equals the impedance. So the Voltage Probe in the simulation yields impedance directly. Using an AC current source lets SPICE sweep across frequency. The "trick" is to set the simulation to be "AC Sweep." For this simulation, run it from 1 kHz to 1 GHz. The 100 Megohm resistor is a practical necessity; SPICE treats a capacitor as a short circuit at AC and so the resistor helps it to simulate the capacitor correctly. Both of these shortcuts (using the 1 A current source and using a high-ohm resistor) can be used later, such as in the chapter on power-supply design.

The results are shown in Fig. 3.5.

To get this result, be sure both the X-axis and Y-axis are set to a logarithmic scale. The SPICE results match the theory, showing the capacitive, resistive, and inductive behavior of an actual capacitor, and the simulation is correct and accurate out to 10's of GHz of frequency.

This leads to the second method for modeling components, which is to draw on knowledge of how the devices and structures behave. The knowledge may come from datasheets, from measuring a sample, or from experience.

The third method is to perform detailed analysis using advanced simulation software, such as 2D and 3D field solvers. For complicated, custom-made structures, like antennas or jet fighters, there is no substitute for this step.

The Limits of Modeling

Modeling and simulation have very important limitations, and the results of modeling always need to be understood inside these limitations.

The first is the accuracy of the component values and parameters of the simulation. To what extent are they known accurately? To what extent are they educated guesses?

The second is the fact that, in electronics, there are tolerances and drift to be considered. Tolerances are a reference to the fact that no two parts are exactly alike. For example, a typical surface-mount resistor tolerance is 1%, meaning that resistors have the stated value plus or minus 1%. Capacitors are usually much wider in tolerance. Drift is a reference to the change in the value of components over time. You can run a simulation, but it only describes the performance of a single set of parameters. What happens if you make 1,000,000 of them? And what will they be like 5 years from now? The way to model this is to perform *sensitivity analysis*, in which the variation in output as a function of changes in circuit parameters is measured. If the circuit behavior changes dramatically with a small change in value, then there will be a problem in mass production and over time.

The third is the fact that some of the estimated values, like resistance, may change according to the state of the circuit. Generally speaking, resistance is frequency-dependent (due to dielectric loss and the skin effect, more on which later). Capacitances inside transistors are voltage-dependent. The model may have to incorporate these effects for maximum accuracy, but at the cost of a considerably more complex model. This is the advantage of using simulators like SPICE – it can take into account frequency-dependent aspects of a model, such as dielectric loss.

The result is that modeling can be extremely misleading, unless great care is taken to put in accurate parameters and the results are understood inside the limits of the model.

Appendix

1. Consider an output from a chip that is driving a circuit with 50 pF of capacitance. The output switches between 0 and 3.3 Volts, and toggles at an average frequency of 625 MHz.

   (a) How much power does it dissipate?
   (b) Where is the power dissipated – in the chip, the wiring, or the circuit that the chip is driving?
   (c) What three steps could be taken to reduce power dissipation?

2. A component has a rated maximum temperature of 100 °C and a thermal coefficient of 15 °C/W.

   (a) At an ambient of 40 °C and a power dissipation of 5 Watts, is the maximum temperature exceeded?
   (b) What is the maximum power dissipation at 40 °C?
   (c) An engineer wants to use the component in a design with a maximum ambient temperature of 50 °C and a power dissipation of 5 Watts. What thermal coefficient would be needed?
   (d) What can be added to a component to reduce its thermal coefficient?

3. Consider the design of a microprocessor. It has a core voltage of 5 V, requires 100 mA of current, and operates at a maximum ambient temperature of 40 °C.

   (a) With a thermal coefficient of 80 °C/Watt, how hot does the microprocessor get? $Tj = 40 + (0.1 \times 5) \times 80 = 40 + 40 = 80$ °C.
   (b) At what ambient temperature would the microprocessor reach 150 °C? Can get 70 °C hotter, so can operate up to $40 + 70 = 110$ °C.

4. Consider a 78 M05 voltage regulator design:

   - The regulator goes into thermal shutdown at a junction temperature of 125 °C or above.
   - 1 Amp output current
   - The regulator takes in voltage at 9VDC and outputs 5VDC. So the power that it dissipates is calculated by multiplying the current by 4 Volts (the voltage drop).
   - Heatsink design has a thermal coefficient of 45 °C/Watt. (The junction temperature is 45 °C/Watt above the ambient temperature.)

     (a) At what ambient temperature does the 78 M05 go into thermal shutdown?
     (b) Consider the same design, except that now it operates at a maximum ambient of 80 °C. At what current does it go into thermal shutdown?

5. Consider a capacitor has a capacitance of 10 µF, a resistance of 0.2 Ω, and an inductance of 0.3 nH.

(a) Calculate the self-resonant frequency. (See Eq. 3.7.)
(b) Construct a SPICE simulation and compare results. The SPICE simulation should display the impedance of the circuit over frequency on a log-log scale.

6. A datasheet lists the self-resonant frequency for a 25 µF capacitor as 5.2 MHz. What is the ESL of the capacitor?

Chapter 4
Circuit Elements: Resistance, Capacitance, and Inductance

Background and Objectives

This chapter reviews the concepts of resistance, capacitance, and inductance in depth. Even though most electrical and computer engineers have studied and used these concepts for some time, the concepts are often misunderstood. When you are finished with this chapter, you should be able to:

- Explain what resistance is
- Use resistivity to calculate resistances of common structures
- Explain what capacitance is
- Explain the origin and physical significance of $I = C dV/dt$
- Explain what inductance is
- Explain $E = L dI/dt$

Reviewing the Review

Back up in Chap. 1, we studied basic concepts like charge (quantity of electrons), current (rate of electron flow), and voltage (energy per electron). Two other important concepts are that of an E field (measured in volts per meter) and that of a B field (a magnetic field associated with current flow).

These quantities are associated with three fundamental circuit parameters, resistance (R), capacitance (C), and inductance (L). Circuit elements that manifest one of these parameters are considered passive, because they only dissipate energy. The associated physical phenomena are generally considered *parasitic*, as they tend to remove energy from a system, but can be exploited cleverly. For example, capacitors are used to store information, and inductors can be used to create very high voltage spikes. Understanding them is critical.

© Springer Nature Switzerland AG 2022
S. H. Russ, *Signal Integrity*, https://doi.org/10.1007/978-3-030-86927-4_4

If one applies a voltage across a piece of material, there may be a measurable flow of current. If not, we classify the material as an *insulator* and, if so, a *conductor*. Of course, some materials have electrically controllable current flow, and we call these *semiconductors*, but that is outside the scope for the moment.

Applying a voltage across a piece of material results in one of two possible outcomes, a flow of current or a static voltage (a voltage with no accompanying current flow). If current flows, meaning the material is a conductor, then the arrangement forms a resistor (and possibly a capacitor). If no current flows, meaning the material is an insulator, then the arrangement forms a capacitor.

What Is Resistance?

Consider a rectangular slab of conductive material of length ℓ (lower-case L), width w, and thickness t. Connect a voltage V across the length of the slab. This is illustrated below in Fig. 4.1. The observation can be made that the current I (flowing through the slab) is proportional to the applied voltage. Also note that the application of a voltage V across the slab results in an E field of size V/ℓ.

If the cross-sectional area (in the example, the product of w and t) is doubled and the voltage stays the same, the amount of current doubles. However, doubling the cross-sectional area doubles the current-carrying area, and so the current density (Amps per unit of cross-sectional area) remains the same. This is a crucial observation that is often overlooked.

If the length is doubled, the current flow is cut in half. However, keeping the voltage the same and doubling the length also cuts the applied E field (applied voltage divided by length) in half. So the following observation can be made:

Given a rectangular slab of material, the ratio of applied E field to current density J is a constant for that material.

Fig. 4.1 Example of a conductor

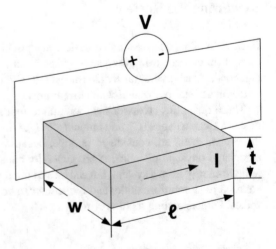

[We call this material constant *resistivity* and give it the symbol ρ (Greek lower-case letter rho). In other words,

$$\rho = E/_J \tag{4.1}$$

where $E = E$ field strength (in V/m) and $J =$ current density (in A/m^2).

What, exactly, is going on here? There are two ways of looking at it. From one perspective, the more voltage you apply, the more energy each electron has and so more electrons want to flow through the conductor. They are more motivated. From another perspective, as the electrons flow through the conductor, imperfections in the conductor cause the electrons to lose energy. An electron losing energy is, by definition, a voltage drop. So as electrons flow, they lose voltage. It is a process very much like friction. Either way of looking at it, a current flowing through a conductor loses voltage, and if you increase the voltage, you get more current.

The units are confusing and become extremely important. The actual unit for E field is volts per length of conductor. The actual unit for current density J is amps per cross-sectional area of conductor. Defining conductor length as ℓ and conductor cross-sectional area as a, we have

$$\rho = E/_J = \frac{V/_\ell}{I/_a} = \frac{Va}{I\ell} \tag{4.2}$$

Recall from introductory electrical engineering that the quotient of voltage divided by current is termed electrical resistance and has units of ohms. Applying unit analysis to resistivity, it has units of ohms times square meters divided by meters, and so has units of ohm-meters. But this is misleading because the "meters" term is actually cross-sectional area divided by length. So resistivity actually has units of ohms-times-area-per-length. It is important to keep the area and the length in the definition because they represent the cross-sectional area and length, respectively, of the resistor.

Each material has its own value of resistivity. For example, the resistivity of silver (the most conductive metal) is $1.59 \times 10^{-8} \Omega \cdot m$. Copper, the most commonly used conductive material, weighs in at $1.68 \times 10^{-8} \Omega \cdot m$. So it turns out copper is almost as conductive as silver, and much less expensive. Iron is $1.0 \times 10^{-7} \Omega \cdot m$ and so is much more resistive (almost a factor of 6).

A review of concepts is in order. Every material has an intrinsic property called resistivity, which is the ratio of applied E field to current density. This creates the well-known phenomenon that the voltage across a piece of material is proportional to the current flow through it. If the voltage is kept the same and the length is doubled, the current flow is cut in half. If the voltage is kept the same and the area is doubled, the current flow is doubled.

We can take one more step, and define the ratio of voltage to current as *resistance*. We do this mainly as a convenience because the conductors we encounter in daily life have a fixed length and area.

For a conductor with a fixed-length ℓ and cross-sectional area a, one can calculate a term that is a function of its resistivity, length, and area. We call this property *resistance* or *R*. Rearranging Eq. 4.2, we have

$$\frac{V}{I} = \rho \frac{\ell}{a} \equiv R \qquad (4.3)$$

In other words, $V=IR$. Now we can calculate resistances. . .

Example 4.1

Consider a bond wire, a wire that connects a signal of an integrated circuit to its package. Bond wires are typically 1 mil (1/1000th of an inch) in diameter and are 0.1″ long. They are made of gold (because gold is malleable and so can be extruded) which has a resistivity of approximately $2.5 \cdot 10^{-8}$ Ωm.

(a) What is the resistance of the bond wire?

The first step is to convert everything to MKS units. The wire diameter is $0.001″ = 0.0254$ mm, and so the cross-sectional area is $3.14 \times (0.0254/2)^2 = 5.067 \times 10^{-4}$ mm^2 $= 5.067 \times 10^{-10}$ m^2. The length is $0.1″ = 2.54$ mm $= 0.00254$ m.

Applying Eq. 4.3, we have $R = 2.5 \times 10^{-8} \times 0.00254/5.067 \times 10^{-10} = 0.125$ Ω.

(b) Sometimes to lower resistance on power connections, two bond wires are used. What is the resistance in this case?

Since the number of wires doubles, the cross-sectional area also doubles and the resistance is cut in half. So $R = 0.125/2 = 0.0625$ Ω.

There are two other ways of expressing resistance that is useful.

The first alternate expression of resistance has to do with wiring, in which resistance is expressed in terms of resistance per unit length. This is convenient because one can simply multiply by the length of an actual piece of wire to determine the actual resistance.

Example 4.2

(a) What is the resistance per unit length of 20-gauge copper wire?

To answer this, we start by delving into the definition of 20 gauge. This is an archaic unit of measure. "Gauge" is shorthand for "American Wire Gauge" and that is why the side of 20-gauge wire is often labeled AWG 20. (That also lets you read the wire gauge of the extension cord one needs for festive holiday lights, an important life skill for electrical engineers that don't want to burn down their houses over the holidays.)

To make a long story short, 20-gauge wire has a diameter of 0.032″ or 0.8128 mm. Its cross-sectional area, therefore, is $\pi \times (0.8128/2)^2$ mm^2 $= 0.5189$ mm^2 $= 5.189 \times 10^{-7}$ m^2.

Recall that we are calculating resistance per length or R/ℓ. Rearranging Eq. 4.3, we have that $R/\ell = \rho/a$. Since the resistivity of copper is $1.68 \times 10^{-8}\Omega \cdot m$, we have that $R/\ell = 1.68 \times 10^{-8}/5.189 \times 10^{-7} = 0.0323 \ \Omega/m = 32.3 \ m\Omega/m$.

(b) What is the resistance of 0.3 m (about 1 foot) of 20-gauge wire?

$32.3 \ m\Omega/m \times 0.3 \ m = 9.69 \ m\Omega$

For conductors of constant cross-section, then, it is possible to calculate resistance per unit length.

The second alternate expression of resistance is that of materials of constant thickness. This frequently arises in circuit boards and inside integrated circuits, for example, because power (and ground) is distributed using a plane or sheet of conductive material.

If you go all the way back up to Fig. 4.1 and consider a case in which the thickness (labeled h in the diagram) is held constant, an interesting property arises. For a material of known or fixed thickness t, the resistance is constrained by the length ℓ and the width w. As ℓ gets longer resistance goes up and as w gets wider resistance goes down. R is a function of the ratio of length to width. Starting with Eq. 4.3 and substituting $t \times w$ for area, resistance can be expressed as

$$\frac{V}{I} = \rho \frac{\ell}{t \times w} = \frac{\rho}{t} \times \frac{\ell}{w} \equiv R \tag{4.4}$$

This type of resistance, then, is a function of the *ratio* of length to width, not the length or width itself. If the length is doubled and the width is doubled, the resistance stays the same. This type of resistance is called *sheet resistance*.

The units of sheet resistance are confusing. The resistance of a piece of material is the sheet resistance times the length and divided by the width (i.e., resistance equals sheet resistance times the length-to-width ratio). The units of length and width cancel and so the units are just ohms. To make clear that sheet resistance is being discussed, and therefore sheet resistance needs to be multiplied by ℓ/w, the units are referred to as "ohms per square." $1 \ \Omega$ per square means that a square piece of material will have a resistance of $1 \ \Omega$. Remember the weird property of sheet resistance that the size of the square does not matter, only the ratio of length to width.

Example 4.3
(a) What is the sheet resistance of 1-ounce copper?

Unfortunately, the thickness of copper on a circuit board is still called out in archaic units. 1-ounce copper gets its name from 1 ounce (weight) of copper per square foot of circuit board. 1-ounce copper is 1.37 mils (1.37 thousandths of an inch) or about 0.035 mm (35 μm) thick.

Sheet resistance $= \rho/t = 1.68 \times 10^{-8}\Omega \cdot m/35 \times 10^{-6} \ m = 0.00048 \ \Omega = 0.5 \ m\Omega$.

(b) What is the resistance per length of a 5-mil-wide trace?

5 mils $= 0.005'' = 0.127$ mm. 1 cm of the trace has $\ell/w = 10/0.127 = 78.7$.
$0.5 \ m\Omega \times 78.7 = 39.37 \ m\Omega$. So the resistance per length is about 40 mΩ/cm or
about 100 mΩ/inch. This is a very handy rule of thumb to remember: A 6-inch 5-mil
circuit trace, for example, has about 0.6 Ω of resistance.

What Is Capacitance?

If two points in space are separated by an insulating material, and if a voltage
difference is applied between the two conductive points, then an electric charge
will accumulate. The observation can be made that, for a fixed arrangement of
conductive points and insulator, *the ratio of charge Q to voltage V is constant.* We
call this ratio *capacitance* and so we have

$$C \equiv Q/V \tag{4.5}$$

Electric charge Q is in units of Coulombs (C). (1 Coulomb of electrons is
6.241×10^{18} electrons.) Voltage V is in units of Volts. Capacitance is in units of
Farads (F). $1F = 1 \ C/1 \ V$.

It is easy to take the derivative of Eq. 4.5 and find out that $I = C \, dV/dt$, but no one
really learns much from doing it, so let's approach it a little differently.

A capacitor is a bucket for storing charge. We frequently use terminology like
"the 1 μF capacitor was charged to 12 V." This is correct, as long as one respects
the fact that the capacitor actually held an amount of charge equal to the product of
the voltage and the capacitor's capacitance value. In this example, it is 12 μC. The
bucket analogy is illustrated further in Fig. 4.2.

Consider the four capacitors illustrated in Fig. 4.2. The height of the charge-
storing bucket is equal to the voltage. That is, the capacitor is a voltage-controlled
charge-storing bucket. The height of the walls of the bucket is proportional to the
voltage. So capacitor 1, with a high voltage, stores more charge and capacitor 2, with
a lower voltage, stores less charge.

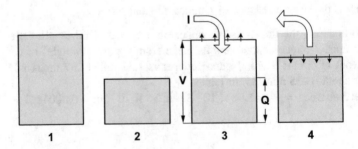

Fig. 4.2 A capacitor is like a bucket for storing charge

Consider capacitor 3. Its voltage is being raised, so the sides of the bucket go up and more charge flows in. If you connect a voltmeter and ammeter to a real capacitor, you can see this – turning up the voltage causes a brief spike in positive current (current flowing into the passive element, by convention) until the capacitor reaches its new charge.

Consider capacitor 4. Its voltage is being lowered, so the sides of the bucket go down and the capacitor dumps out charge. Turning down the voltage causes a brief spike in negative current (current flowing out) until the capacitor reaches its new charge. When operated in this downward-voltage mode, the capacitor sources current instead of sinking it.

To recap, turning down the voltage dumps out charge, and turning down the voltage faster dumps the charge out at a faster rate. So the rate of flow of charge is proportional to the rate of change of voltage. Since the rate of flow of charge is current, current is proportional to the rate of change of voltage. $I = C \, dV/dt$.

Charging a capacitor is more complicated than many realize.

Can current flow through an open circuit? So if turning up the voltage causes current flow, and if the capacitor is an open circuit, how can current flow? How can current flow through an open circuit?

The answers are surprising. What actually happens is this: When the voltage goes up, the positive plate attracts positive charge and the negative plate attracts negative charge. Thus, there is positive current going into the positive plate and negative current going into the negative plate. This has the appearance of positive current going into the positive plate and *positive* current *leaving* the negative plate. In other words, it has the appearance of current flowing in a circle through the capacitor. What is actually happening is that charges are redistributing themselves.

There are two additional properties of capacitors that are important to understand.

First, capacitors can store energy. (See Eq. 3.3 and the accompanying discussion for the equation and the details of the calculation.) The energy is actually stored *inside the insulator* between the conductive plates of the capacitor. More specifically, the energy is stored *in the E field* that is created between the plates. This distinction is important – it is easy to confuse the two because the E field happens to lie in the insulator. The reason it is important to remember that the E field stores energy is that the E field can squeeze out past the boundaries of the conductors due to fringing.

Second, a change in the E field between the two conductors does not occur instantaneously. Picture a situation in which the two capacitor's plates are very long wires with an insulator between them. A change in the E field at one end of the capacitor moves at a finite speed down the length of the conductors. Changes in the E field propagate at a finite speed, equal to the speed of light divided by the square root of the dielectric constant of the insulator. The changing E field will race down the length of the wires (the length of the capacitor's conductors) at that speed. This is another way of saying that your selfie moves at the speed of light when it travels from your phone to the cell phone antenna: It moves in the changing E field between the phone and the antenna. The relative dielectric constant of the air is nearly one, and so it moves at a speed very close to the speed of light.

Finding Capacitance

So how do we calculate the value of capacitance?

Let's reconsider our capacitor, but with a special shape in mind. Consider two flat plates, each with area A, separated by an insulating material, and held a height h apart. (For this to work, the plates have to be much larger than the separation between them, for reasons to be explained below.) If a certain voltage V is applied between the plates is applied, there is an accumulation of charge Q. This is illustrated below in Fig. 4.3.

If the h increases, but V and A are held constant, the amount of charge stored actually drops. Think of it this way: As the h increases, the opposite charges are less eager to line up on the plates. A more technically correct way of looking at it is this: The amount of charge stored is proportional to the E field between the plates, which is, in turn, proportional to V/h. In other words, Q is linearly proportional to the strength of the E field, and the strength of the E field is inversely proportional to h.

If area A increases but V and h are held constant, the amount of charge Q stored increases linearly with A.

To recap, Q is proportional to A and proportional to the E field between the plates.

The ratio of charge stored to E field is called the *dielectric constant*, for which the symbol is ε (Greek lowercase epsilon). So we can write the equation

$$Q \approx \varepsilon AE = \varepsilon A V/h \tag{4.6}$$

The dielectric constant is an intrinsic property of the insulating material. If the insulator is a vacuum, there is still some ability to store charge (i.e., nonzero charge accumulates if a voltage is applied), and it is called the dielectric constant of free space or ε_0 (pronounced "epsilon-nought"). If the insulator is anything else, it has a higher ability to store charge. We express this ability as a multiple of ε_0 and so we say that $\varepsilon = \varepsilon_r \varepsilon_0$, where ε_r is the *relative dielectric constant* of the material. For example, the FR-4 fiberglass used to make most circuit boards has $\varepsilon_r = 4.2$.

Note that Eq. 4.6 is not an equality. It is an approximation that becomes more accurate as A gets larger and h gets smaller. The inequality occurs because the E field

Fig. 4.3 Example of a capacitor. Note how the E field extends past the ends of the plates

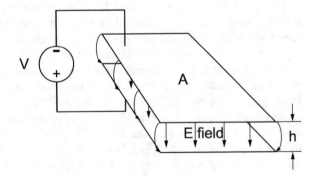

is not a strict linear function of h. Rather, the E field is formed between the areas of opposite charge. Past the ends of the plates, the E field bows out into the surrounding space (an effect called *fringing*), and so the E field is actually larger than the volume of space formed directly between the plates. So to be completely precise, the ratio of charge to voltage is governed by the E field, not by V/h. That is why people who study electromagnetics speak in terms of the E field storing a charge.

Overlooking this for the moment, and substituting C for Q/V and substituting $\varepsilon_r\varepsilon_0$ for ε, Eq. 4.6 can be rewritten as

$$Q/V = C \approx \varepsilon A/h = \varepsilon_r\varepsilon_0 A/h \qquad (4.7)$$

This is called the *parallel plate capacitor equation*. Many people forget that this equation is an approximation. The actual amount of capacitance between parallel plates is larger than that predicted by the equation because, as noted above, the E field fringes out past the ends of the plates.

Fringing has other effects as well. Since the E field can extend past the conductors, it means that the E field can "reach out" and wreak havoc at a distance. For example, if the conductors are a power and ground plane that extend all the way up to the edge of a circuit board, the E field can extend outside the circuit board and into the chassis holding the board. This fringing E field is one of the more common sources of signal-integrity headaches.

As with resistances, the units for $\varepsilon = \varepsilon_r\varepsilon_0$ can be confusing. First, ε_r is a dimensionless ratio. Second, the units for ε_0 have to be in units of capacitance times height divided by area, and so have MKS units of F/m. As with resistance, this is misleading. It is actually units of Farads-times-height-per-area.

Example 4.4

(a) What is the capacitance per unit area on a circuit board?

Rearranging Eq. 4.6 to solve for C/A, we obtain that $C/A = \varepsilon_r\varepsilon_0/h$. Since $\varepsilon_r = 4.2$ and since $\varepsilon_0 = 8.854$ pF/m, we have that $C/A = (4.2 \times 8.854 \text{ pF/m})/h = 37.19/h$ pF/ m^2 for h in meters.

(b) For a plane spacing of 1 mm, what is the capacitance between the power and ground plane per area of a 100 cm^2 circuit board?

Since $h = 0.001$ m, $C/A = 37.19/(0.001) = 37.19 \times 10^3$ pF/m$^2 = 37.19$ nF/m^2. Converting to cm^2, this corresponds to 37.19 nF/$10^4 = 3.7$ pF/cm^2. So if a circuit board has an area of 100 cm^2, the power and ground plane form a 3700 pF capacitor.

So what about signals on a circuit board? The equations are quite complicated, but fortunately, IPC (the standards-making body for the circuit-board industry) has published them.

The first equation is for stripline, a configuration in which the signal is on the top layer and is routed over a ground plane. As illustrated below, for conductor width w and thickness t, located h over the ground plane, the equation is as shown in Fig. 4.4.

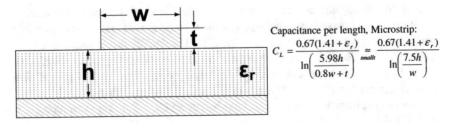

Capacitance per length, Microstrip:

$$C_L = \frac{0.67(1.41+\varepsilon_r)}{\ln\left(\dfrac{5.98h}{0.8w+t}\right)} \underset{small\,t}{\approx} \frac{0.67(1.41+\varepsilon_r)}{\ln\left(\dfrac{7.5h}{w}\right)}$$

Fig. 4.4 IPC equation for stripline capacitance

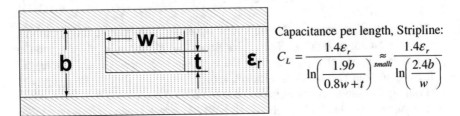

Capacitance per length, Stripline:

$$C_L = \frac{1.4\varepsilon_r}{\ln\left(\dfrac{1.9b}{0.8w+t}\right)} \underset{small\,t}{\approx} \frac{1.4\varepsilon_r}{\ln\left(\dfrac{2.4b}{w}\right)}$$

Fig. 4.5 IPC equation for microstrip capacitance

Note that h, w, and t are in mils and C_L is in pF/inch. Also note the second equation is for the situation in which t is much smaller than h. (As is common in the circuit-board industry, Imperial units are used instead of metric units.)

The second equation is for microstrip, a configuration in which the signal is buried between two planes. For a conductor of width w and thickness t, located exactly halfway between planes that are b apart, the capacitance per length is calculated as shown in Fig. 4.5.

Note that b, w, and t are in mils and C_L is in pF/inch.

Both of these equations are approximations, and break down if w or t are relatively large. They break down if the signal is located close to the edge of the ground plane (and therefore close to the edge of the circuit board also) because the E field fringes outside the circuit board. In the case of stripline, the equation is less accurate if there is solder mask. The actual capacitance is higher than the equation's prediction in this case because solder mask is a better insulator than air, can store more charge, and drives up the capacitance. In the case of microstrip, the equation only works if the signal is centered vertically between the planes; this only rarely happens in practice.

If you need more accuracy, you need a 2D field solver to find capacitance per unit length. You need a 3D solver if the traces make right-angle turns or have nonconstant cross-section.

What Is Inductance?

Every electrical and computer engineer knows that $E = L\, dI/dt$, but really where does that equation come from? What, exactly, is L? To figure this out, we have to start from the beginning.

Around any current flow, there are circular magnetic-field loops. The fact that moving electrons create a magnetic field is due to relativity (long story), but the point is that if there is a flowing electric current, there is a magnetic field around it, and the amount of magnetic flux linearly is proportional to the current. The flow of current and corresponding magnetic-field lines is illustrated below in Fig. 4.6.

Magnetic flux is measured in Webers. Webers is one of those arbitrary numbers, like Coulombs, and represents a fixed, arbitrary amount of magnetic flux. Inductance is simply the ratio of magnetic flux to current. A current of 1 Amp producing 1 Weber of magnetic flux corresponds to 1 Henry of inductance. For a magnetic flux B and a current I, the inductance L is defined as

$$L = B/I \tag{4.8}$$

In other words, magnetic flux B is equal to the inductance L times current I.

The relationship between current and magnetic flux is illustrated below in Fig. 4.7.

On the left, a moving electron with its magnetic field is shown. In the middle, more electrons equal more magnetic fields. In other words, the amount of magnetic flux is linearly proportional to the current because more current equals more moving electrons equals more magnetic flux. On the right, electrons moving down a longer wire is illustrated. The amount of magnetic flux per electron is linearly proportional

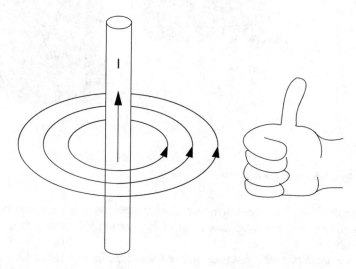

Fig. 4.6 Current and magnetic field. Note how the field follows the right-hand rule

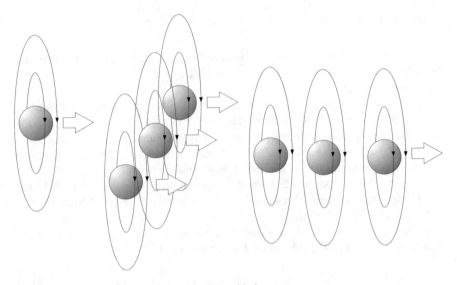

Fig. 4.7 Relation of current and magnetic flux and inductance

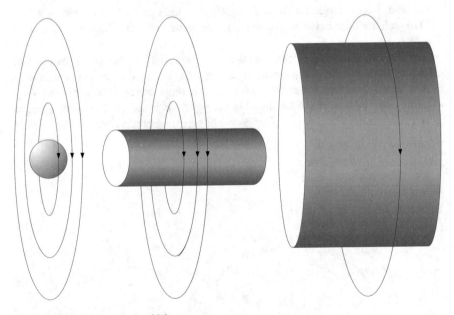

Fig. 4.8 Inductance and wire thickness

to the length of the wire because a longer wire has more room for more magnetic flux lines. In other words, the inductance (ratio of flux to current) increases linearly with wire length.

The amount of magnetic flux goes up as the wire gets thinner. This is illustrated in Fig. 4.8.

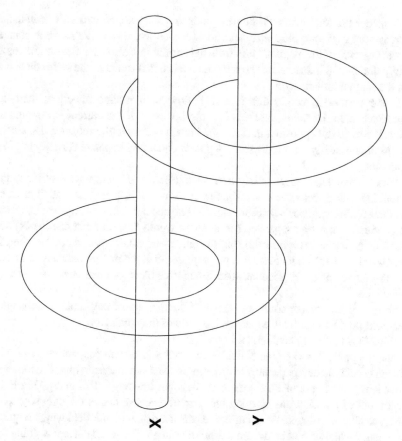

Fig. 4.9 Mutual and self-inductance

On the left, a single electron with magnetic-field lines is shown. In the middle, a thin wire has more of the flux external to the wire. There is less room inside the wire to contain the magnetic flux, and so more "leaks out" of the wire into free space. On the right, a thicker wire is shown. More of the magnetic flux is confined inside the wire. The result of all this is that a thinner wire has more inductance – for the same amount of current, more flux is located externally to the wire.

The amount of magnetic flux is also a function of the material the wire is made of, specifically if the material is ferromagnetic or not. The inductance with respect to materials is not unlike that of capacitance. Free space has a certain, nonzero permeability called μ_0. The permeability of other materials is usually expressed as a ratio relative to μ_0, and the ratio is called μ_r or the relative permeability. The relative permeability of most materials, including copper and air, is nearly 1. The relative permeability of iron is around 5000.

If the magnetic flux is caused by current in the same conductor, it is called *self-inductance* (L_{self}). If the magnetic flux is caused by current in a neighboring conductor, it is called *mutual inductance* (L_{mut}). This is illustrated below in Fig. 4.9.

Consider the total magnetic flux in conductor X. First, the current in conductor X creates its own magnetic flux, and, as noted above, this is the self-inductance. However, some of the magnetic flux from the current in Y also encircles conductor X. This is the mutual inductance. The total inductance is the sum of the self-inductance and the mutual inductance.

If the current in conductors X and Y flow in the same direction, then the inductances add. However, and this is a big however, if the current in conductors X and Y flow in the opposite direction, the inductances *actually subtract*. This is the way to obtain nearly zero inductance – put an equal and opposite current next to a conductor.

Mutual inductance must satisfy two rules. First, it is always symmetric, so the mutual inductance between X and Y is the same in X as it is in Y. Second, it must be less than each conductor's self-inductance. The second rule is always true because the conductors must be separated (or else you would have one big conductor) and therefore there is always some flux that surrounds one conductor and not the other. In other words, the self-inductance is always bigger than the mutual inductance. If you take these two rules into account, then it becomes clear you can never obtain zero inductance.

So how does voltage come into play? If the amount of magnetic flux changes, there will be a voltage induced across the ends of the conductor.

What is actually happening is this:

The magnetic flux (i.e., the B field) surrounding a current can store energy, just like the E field inside a capacitor. In order to increase the magnetic flux, extra energy is needed. An increase in B field requires the input of energy. This energy has to be sucked out of the electrons carrying the current, because that is the only source of energy available. An increase in current results in an increased B field which requires real energy which is obtained from a drop in voltage. While the current is going up, there is a measurable drop in voltage.

Conversely, decreasing the magnetic flux causes a rise in voltage because the decreasing magnetic field is dumping its energy into the passing electrons. The energy that is released from the decrease in magnetic flux is being dumped into the electrons carrying the current. (It has to be dumped somewhere, because of conservation of energy, and so they are dumped into the passing electrons.) A decrease in current results in a reduced B field which releases real energy which increases the voltage of passing electrons.

This behavior (both the increase and decrease) is called Faraday's Law: When the magnetic flux linking a circuit changes, the voltage in the circuit changes linearly with respect to the rate of change of magnetic flux.

A rapid decrease in current will cause an inductor to emit a large spike in voltage. This is how spark plugs worked in older cars; the distributor cap broke the circuit that was pushing current through a coil, and the voltage spike was high enough to create a spark inside an engine's piston. Circuits that drive inductive loads usually require some form of protection to handle this voltage spike, such as a diode in parallel with the inductor.

As noted above, the amount of voltage change is proportional to the rate of change of magnetic flux. Roughly speaking, if the magnetic field is dropping faster, more energy is dumped into each electron. Noting that magnetic flux is equal to current times inductance (by definition), we have

$$V = {\Delta B}/{\Delta t} = {L\Delta I}/{\Delta t} = L{dI}/{dt} \qquad (4.9)$$

To restate, Eq. 4.9 is a result of the facts that the magnetic flux stores energy and that the energy is either obtained from or dumped into the moving electrons if and when the flux changes.

So what happens if current in a neighboring conductor changes? For example, earlier in Fig. 4.5, what happens in conductor X if the current in conductor Y changes? A voltage is induced in conductor X! The voltage in conductor X, $V_x = L_{mut} \, dI/dt$. This is the source of a great deal of signal-integrity headaches, L_{mut} and dI/dt.

Calculating Inductance

Inductance is famously hard to calculate. Inductance is actually defined as the total inductance around the entire current loop, and so has the odd property that an instrument making an inductance measurement is necessarily part of the loop. To be accurate, the inductance of the entire loop of current has to be summed up.

Faced with this dilemma, most speak of partial inductance, which is the inductance of pieces of the circuit. Calculating partial inductance (the inductance of part of the loop) turns out to be useful, because it at least permits comparisons. For example, one can calculate the partial inductance of a through-hole pin and compare it to the partial inductance of a surface-mount lead.

The self-inductance L_{self} of a round rod is found as

$$L_{self} = 5d \left[\ln \left(\frac{2d}{r} \right) - \frac{3}{4} \right] \qquad (4.10)$$

where L = Inductance (nH), d = *length* in inches (NOT diameter), and r = radius in inches.

The mutual inductance L_{mut} of two neighboring round rods is found as

$$L_{mut} = 5d \left[\ln \left(\frac{2d}{s} \right) - 1 \right] \qquad (4.11)$$

where L = Inductance (nH), d = *length* in inches (NOT diameter), and s = center-to-center spacing in inches. This equation assumes that $s \ll d$ (i.e., that two rods are much closer together than their length). As with capacitance, these equations are approximations.

Example 4.5
A bond wire is the wire that is used to connect an integrated-circuit pad to the package pin that houses the integrated circuit. In other words, a bond wire connects a chip to the outside world. Consider two adjacent bond wires. The wires are 0.1 inches long and 1 mil in diameter, and are spaced 5 mils apart.

(a) What is the self-inductance in one bond wire?

$L_{self} = 5 \times 0.1 \times (\ln\{2 \times 0.1/0.0005\} - 3/4) = 2.6$ nH.

(b) What is the mutual inductance between two bond wires?

$L_{mut} = 5 \times 0.1 \times (\ln\{2 \times 0.1/0.005\} - 1) = 1.3$ nH.

So the mutual inductance between the wires is roughly half each wire's self-inductance. Whether the inductances add or subtract depends on the direction of current flow.

Inductance and Return Current

Recall the switching gate discussed earlier (see Fig. 3.1). When a logic output goes from a logic-1 state (Vdd volts) to a logic-0 state (0 V), the current required to discharge the capacitance (formed by the wire and any logic input) flows in a loop through the logic output *and the ground connection*. The current flowing through the ground connection is called *return current*.

How much inductance does the switching signal see? It all depends! The return current is always equal in magnitude and opposite in direction to the signal current. (This is because the signal current is on the same loop as the return current, and current in a loop is constant.) The mutual inductance between the signal current and return current will therefore always tend to subtract. The total inductance is minimized, therefore, *when the spacing between the signal current and return current is minimized*. Thus, the return current tends to be located on the ground plane directly below the signal current, as illustrated below in Fig. 4.10.

Figure 4.10 is a zoomed-in view of a signal trace (a wire on a circuit board) over the ground plane. The figure is drawn as if the circuit-board insulating material is transparent; only the signal trace and ground plane are shown. There is a signal current flowing (to the left) on the signal trace and an equal and opposite return current (to the right) on the ground plane. The return current is in the area shaded in gray; it literally shadows the signal trace on the ground plane. The return current is shown as fuzzy on purpose – it actually spreads out a little wider than the signal conductor to minimize resistance.

The return current specifically does not spread out across the entire ground plane. While spreading out across the plane would indeed minimize resistance, it would

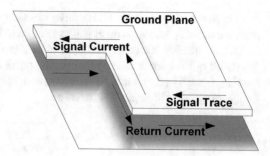

Fig. 4.10 Signal and return current

increase inductance. Since current follows the path of least impedance, not least resistance or inductance, the inductance dominates and return current is found mostly, but not entirely, on the ground plane underneath the signal.

The actual shape of the return current is a compromise between minimal resistance (which would be the current spread evenly across the ground plane) and minimal inductance (which would be the current huddled up underneath the signal conductor.) Again, *impedance* is minimized!

There is an extremely important corollary: Any break in the ground plane causes the return current to be forced away from the signal, causing an increase in inductance. Specifically, moving the return current away from the signal current lowers the mutual inductance between the two. While lowering mutual inductance may sound good, recall that, in this special case, the signal and return current's mutual inductance is being used to *cancel* the inductance. Lowering the mutual inductance lowers the cancellation. The signal "sees" the inductance of the entire loop, including both the inductance of the signal trace and the inductance of the return-current path. Since cancellation is going down, the net inductance is going up. The bottom line is the inductance seen by the signal current goes up.

Breaks in the ground plane, such as slots or holes, are undesirable most of the time. Conversely, if a slot or hole is needed (such as a mounting hole), then signals should be routed away from them.

Inductance and the Skin Effect

Go back and consider the magnetic loops surrounding current again, and this time think about what happens *inside* a wire, as illustrated in Fig. 4.7. The conductor at the center of the wire "sees" more field lines than the conductor at the outside because it "sees" the flux from every part of the conductor. The center of the conductor, therefore, has more inductance than the outside of a conductor! Since AC current follows the path of least impedance, it tends to hug the outside of a conductor – that is where there is less inductance.

Thus, inductance leads to an important effect of electric current, the *skin effect*. At AC, current "hugs" the outside of a conductor. Even at the relatively low AC electric-power frequency of 60 Hz there is a skin effect. Large high-voltage power lines are actually hollow – there is no metal there because there is no current there.

Current density is exponential with radius, but there is a useful approximation: Nearly all of the current is in a region at the outside of a conductor to a depth called the *skin depth*. For a material with resistivity ρ, relative permeability μ_r, operating at frequency f, the skin depth δ is found as

$$\delta = \sqrt{\frac{\rho}{\pi \mu_0 \mu_r f}} \tag{4.12}$$

This is for skin depth in meters, f in MHz, and note that μ_0 is the permeability of free space (1.26 Henries per meter).

Example 4.6
Consider a copper conductor on a circuit board.

(a) What is the skin depth at 1 MHz?

Copper has a μ_r of about 1 (since copper is not magnetic) and a resistivity of $1.68 \times 10^{-8} \Omega \cdot$ m. So $\delta =$ sqrt($1.68 \times 10^{-8} \Omega \cdot$ m/ ($\pi \times 1 \times 1.26 \times 10^{-6} \times 1$ MHz)) $= 65$ μm.

(b) At what frequency does the skin depth fall below 35 μm, the thickness of 1-ounce copper?

Rearranging the equation, we have $f = 1.68 \times 10^{-8} \Omega \cdot$ m/ ($\pi \times 1 \times 1.26 \times 10^{-6} \times (35$ μm)^2). So $f = 3.4$ MHz. In other words, above 3.4 MHz, the signal current no longer fills the entire conductor; it begins to hug the side of the conductor facing the ground plane.

(c) What rise time corresponds to a knee frequency of 3.4 MHz?

$1/2t_r = 3.4$ MHz. So $t_r = 1/6.8$ MHz $= 14.7$ ns. So signals with a rise time faster than 14.7 ns do not use the entire copper conductor.

Since the skin depth decreases with the square root of frequency, and since the resistance of a conductor increases linearly with decreases in conductor thickness, the apparent resistance of a conductor increases as the square root of frequency. This effect "kicks in" once the frequency of the signal is high enough such that the skin depth is less than the thickness of the conductor. That is, it begins to be observed once the frequency of the signal is so high that it no longer uses the entire conductor. When conductors are operated in this region, called the skin-effect region, the apparent resistance of the conductor goes up with frequency. This leads to the property that a signal "sees" different amounts of resistance at different frequencies or, in other words, sees a frequency-dependent attenuation. Digital signals routed using 1-ounce copper with a rise time faster than 14.7 ns are in this region. Since digital signals have had a rise time faster than 14.6 ns since roughly World War II, and since almost all circuit boards use 1-ounce copper, essentially all digital signals are in this skin-effect region.

Appendix

1. Consider the resistivity (ρ) of the following materials:

Material	Resistivity (Ω-m)
Silver	1.59×10^{-8}
Copper	1.68×10^{-8}
Gold	2.44×10^{-8}
Aluminum	2.82×10^{-8}
Iron	1.00×10^{-7}

   Find the resistance per unit length of 20-gauge wire (0.8128 mm diameter) made from each material.

2. Use the resistivity of the materials listed above to determine the voltage loss of the following wiring scenarios.

   (a) 2/0 AWG wire (9.266 mm diameter), 100 m, carrying 200 Amps: Compare copper to aluminum. (This is the type of wire used to carry electricity into homes in North America.)

   (b) 14 AWG wire (1.628 mm diameter), 10 meters, carrying 15 Amps: Compare copper to aluminum. (This is the type of wire used to carry electricity inside homes in North America.) $R = \rho\ell/A$.
   For Al: $R = 2.82 \times 10^{-8} \times 10/\pi(0.814 \times 10^{-3})^2 = 0.0135 \ \Omega$. $V = 15R = 0.203$ V. For Cu: $R = 1.68 \times 10^{-8} \times 10/\pi(0.814 \times 10^{-3})^2 = 0.00807 \ \Omega$. $V = 15R = 0.121$ V.

   (c) 30 AWG wire (0.255 mm diameter), 0.3 meters, carrying 50 mA: Compare gold to silver.

3. Recall from the text that 1-ounce copper is about 35 μm thick. A company is designing a board with high power requirements.

   (a) What is the sheet resistance of 2-ounce copper?

   (b) What thickness of copper would be needed in order for a 10-mil trace (0.254 mm) that is 15 cm long to have a resistance of 0.1 Ω or less? Assume that copper thickness is only available in integers (1-ounce, 2-ounce, etc.)

4. A company is constructing a circuit board from a material with extremely low dielectric constant, $\varepsilon_r = 1.7$. Consider a circuit trace that is 5 mils wide (0.127 mm) and located 27 mils (0.6858 mm) above a ground plane. The trace is made from 1-ounce copper (35 μm thick).

   (a) Use the parallel-plate capacitor equation to calculate the capacitance of trace that is 10 cm long.

   (b) Use the IPC Microstrip equation to calculate the capacitance of the trace.

(c) Why is the value calculated by the microstrip equation greater? What physical phenomenon does the parallel-plate capacitor ignore?

5. Consider a stripline circuit made from the same materials as the previous problem. In this case, the spacing between ground planes is 40 mils (1.016 mm). As before, the trace is 5 mils wide and 35 μm thick. The trace is centered vertically between the ground planes.

   (a) Use the IPC stripline equation to find the capacitance of 10 cm of the trace.
   (b) Express the capacitance in units of pF/cm.

6. Consider the round-rod equation for self-inductance and for mutual inductance. (See Eqs. 4.7 and 4.8 above.)

   (a) Find the self-inductance for 6 inches (152.4 mm) of 20 AWG wire (0.8128 mm diameter). (This is the size of commonly used "hookup wire" used in electrical engineering lab classes.)
   (b) Find the mutual inductance for 2 6-inch wires with a center-to-center spacing of 2 mm.

7. Consider the equation for skin depth of a material. (See Eq. 4.10.)

   (a) What is the skin depth of gold at 1 MHz?
   (b) At what frequency does the skin depth fall below 12.7 μm, the radius of an integrated-circuit bond wire?
   (c) What rise time corresponds to a knee frequency of the frequency you calculated in part b?

8. A company is constructing a circuit board from alumina (aluminum oxide) with a dielectric constant, $\varepsilon_r = 9.8$. Consider a circuit trace that is 5 mils wide (0.127 mm) and located 20 mils (0.508 mm) above a ground plane. The trace is made from 1-ounce copper (1.4 mils or 35 μm thick).

   (a) Use the parallel-plate capacitor equation to calculate the capacitance of a trace that is 5.08 cm (2 inches) long. $A = 0.127 \times 10^{-3} \times 5.08 \times 10^{-2} = 6.4516 \times 10^{-6} \, \text{m}^2$. $h = 0.508 \times 10^{-3}$ m. $\varepsilon_0 = 8.854 \times 10^{-12}$ F/m. $C = 9.8 \times 8.854 \times 10^{-12} \times 6.4516 \times 10^{-6} \, \text{m}^2 / 0.508 \times 10^{-3} = 1.102$ pF.
   (b) Use the IPC Microstrip equation to calculate the capacitance of the trace. Notice the units of the IPC equation! $C_L = (0.67(1.41 + 9.8))/\ln(5.98 \times 20/(0.8 \times 5 + 1.4)) = 2.42$ pF/inch. 2.42 pF/inch × 2 inches = 2.42 × 2 = 4.84 pF.
   (c) Why is the value calculated by the microstrip equation greater? What physical phenomenon does the parallel-plate capacitor ignore? Fringing.

9. Consider the equation for skin depth of a material.

   (a) Consider R, C, and L. Which one causes the skin-depth phenomenon? L
   (b) What is the skin depth of copper at 50 MHz? Sqrt($1.68 \times 10^{-8}/\pi \times 1.26 \times 10^{-6} \, 50 \times 10^6$) = 9.2 μm

(c) At what frequency does the skin depth fall below 0.127 mm, the radius of 30AWG wire? (For example, this is the type of wire that might be used in a SATA cable.) $f = 1.68 \times 10^{-8}/ (\pi \times 1 \times 1.26 \times 10^{-6} \times (0.127 \text{ mm})^2) = 263 \text{ kHz}$

(d) What rise time corresponds to a knee frequency of the frequency you calculated in part b? $1/(2 \times 263 \times 10^3) = 1.9 \text{ μs}$

10. Consider basic units back from Chap. 1, like E, V, I, Q, B, and J (which is current density, or Amps/m^2).

(a) What are typical units of E? V/m
(b) Express resistivity (ρ) as the ratio of two basic units. E/J
(c) Consider parallel plates of area A. Express the charge Q as the product of permittivity ε, area A, and one of the basic units. εAE
(d) Express inductance L as the ratio of two basic units. B/I

Chapter 5
Ground Bounce and Ringing

Background and Objectives

We have spent the first four chapters studying the basics, including circuit boards and "lumped" parameters such as resistance, capacitance, and inductance. We also studied important consequences of these parameters, such as how the return current follows the signal path. Now we are going to study two of the effects that these lumped parameters cause, ground bounce and ringing. When this chapter is finished, you should be able to:

- Explain the root causes of ground bounce and ringing in terms of resistance, inductance, and capacitance
- Understand and apply formulas that provide approximate estimates of ground bounce and ringing
- Diagnose them by making measurements and/or by observing symptoms
- Describe design practices that minimize ground bounce and ringing
- Describe ways to fix circuits that have problems with ground bounce and ringing

The Role of Inductance

In this chapter, we will still be applying lumped analysis. As we have seen, lumped analysis is only accurate over short distances or at low frequencies, or when the high-frequency part of a circuit or problem can be abstracted away (academic jargon for "ignored"). In both of these cases, you can easily duplicate the results shown here without having to resort to distributed analysis, and the lumped-analysis model provides useful insight into the root physical causes. In other words, we are in a region where lumped analysis works well enough to explain what is going on, and is much easier to understand than distributed analysis.

© Springer Nature Switzerland AG 2022
S. H. Russ, *Signal Integrity*, https://doi.org/10.1007/978-3-030-86927-4_5

Both of the effects we discuss (ground bounce and ringing) are a direct consequence of inductance. If you haven't figured it out by now, inductance is the primary
cause of poor signal integrity; inductance messes everything up.

In the case of ground bounce, we discuss and model the results of inductance in
the power and ground connections that supply power to logic gates. Ground bounce
is quite common because power and ground connections have inductance.

In the case of ringing, the inductance is in the signal connection. Ringing is much
less common today because fast rise times have driven down the critical length and
so signals are much more accurately modeled using distributed analysis than
lumped. We are analyzing briefly here because of its connection to probing signals
on a circuit board, and because there is the chance it can still be encountered in a
low-frequency signal or on a very short signal trace. Also, one has to understand
what ringing is in order to understand how it is fundamentally different from
reflections on a transmission line, even though the two look the same on an
oscilloscope. (Indeed, reflections are often mistakenly referred to as "ringing.")
We will study reflections later.

What Is Ground Bounce?

Consider a logic gate driving a circuit board trace. Since the trace is a conductor over
a ground plane separated by an insulator, it is not only a trace – it is also a capacitor.
The logic input that the gate drives is a CMOS logic structure combined with an ESD
protection device (more on that later) which is highly capacitive. So everything that
the driving gate "sees" is a capacitor, and, if one treats a signal as if it is infinitely
fast, one can accurately model the load of a driving gate as a large single capacitance.
(Again, this is lumped analysis and so we make an "infinitely fast" assumption.)

Now consider what happens when the driving logic gate tries to change the output
from a logic "high" or "1" to a logic "low" or "0." This is shown in Fig. 5.1.

The equivalent needs some explanation. The driving gate looks like a switch that
is being closed (see Fig. 3.1 for comparison). The combination of the signal trace and
the gate that is being driven looks like a single bulk capacitance. In other words, the
capacitance of the trace and logic input can be lumped together into a single unit of

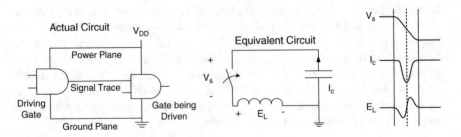

Fig. 5.1 Equivalent circuit and ground-bounce waveform

capacitance called "the load." The driving gate needs to discharge the load, the combined capacitance of the trace, and the logic input.

The driving gate pulls charge out of the load (that is, it discharges the load from a logic 1 to a logic 0) and pumps charge into the ground connection. From the point of view of the driving gate, it is pulling current from the signal trace and dumping current onto its ground connection, the connection between the chip that houses the gate and the circuit board's ground.

However, as we discussed in Chap. 3, that is only half the story. The charge is indeed pumped into "ground," but the charge has to flow in a complete circuit (that is, a complete loop). The "discharged charge" has to flow out of the ground connection of the driving gate, out of the ground connection of the chip in which the driving gate is located, back out onto the circuit board, along the signal that is being discharged, and up into the ground connection of the chip where the input is located. As discussed in Chap. 4, this arrangement of signal and return current minimizes the inductance.

In a world with no inductance, everything would be fine. The driving gate would pull the signal down to a logic low or "0" and the logic input would see a transition from logic 1 to logic 0 once the capacitance had been discharged. However, the inductance (coupled with the capacitance) creates a serious issue.

The driving gate is creating a falling-edge waveform; it is a ramp from logic 1 to logic 0. The load capacitance (trace plus logic input) tends to store charge, and so the falling voltage from the driving gate causes the stored positive charge to "spill out" back to the driving gate (see Fig. 4.2). Use I_c to represent the current through the capacitor and C to represent the load capacitance. Since the current through the capacitor $I_C = C \frac{dV_S}{dt}$, and since V_S (the source voltage) is a falling edge, the current that flows back to the driving gate is the *first derivative of the falling edge*, a downward-facing pulse. (The pulse is downward both because the voltage is falling and because the current is coming into the logic gate which, by convention, is negative.) This current must flow in a loop from the capacitance to the driving gate to the ground connection back to the logic input.

However, and this is a big however, the ground connection itself has inductance. To be more specific, the ground connection that runs from the logic gate, through the chip package, through the package pin that connects to the ground plane, and to the ground plane itself has inductance. This negative spike in current has to traverse the inductance of the ground connection. The change in current causes a voltage across the ground-connection inductance, and the shape of the current waveform doubles the problem, because it is a pulse, a falling edge followed by a rising edge. Since $E_L = L \frac{dI_C}{dt}$, and since I_C is a downward pulse, the voltage across the inductance of the ground network is the derivative of a pulse (and the *second derivative of the falling edge*). It is a double pulse, in this case a downward spike immediately followed by an upward spike. It is big, ugly, and unmistakable. It is the only signal-integrity issue we will discuss in this entire book that can be recognized unmistakably from its oscilloscope image, the double-pulse waveform. It is a direct consequence of driving a capacitive signal through an inductive power and ground connection. It is found in an unexpected place – more on which in a moment – but the shape of the waveform is unmistakable.

(The term "power and ground" is used here because exactly the same discussion occurs on the power rail when the driving gate is trying to create a rising edge; it is a symmetric situation.)

To review, the current I_C needed to discharge the capacitor (in the case of a falling edge) is found by $I_C = C dV_S/dt$ where V_S is the falling-edge waveform. The ground-connection inductance then creates a voltage of $E_L L dI_C/dt = LC d^2V_S/dt^2$. It is L times C times the second derivative of the falling edge.

To make d^2V_S/dt^2 tractable (fancy academic-speak for "solvable"), we have to assume the exact shape of the falling-edge waveform. If one assumes the waveform is a pure decaying exponential, the first and second derivatives are decaying exponentials, which is not the behavior that is observed in the lab. An alternative is to assume that the first derivative (that is, the current waveform) is shaped like a Gaussian "bell curve" or Normal Distribution. The integral of a Gaussian distribution looks like a smoothed-out rising edge, and so the assumption lines up with experience and produces a waveform that can easily be modeled.

Based on the assumption of a Gaussian first derivative, the peak value of the second derivative can be shown to be $1.6/T_r^2$ where T_r is the rise or fall time. And so the estimated ground bounce voltage V_{gb} can be found as

$$V_{gb} \approx \frac{1.6 \Delta V L C n}{T_r^2}, \qquad (5.1)$$

where ΔV is the logic swing (the logic 1 voltage minus the logic 0 voltage), L is the inductance on the ground trace, C is the capacitance being driven by each output, and n is the number of simultaneous switching outputs. Keep in mind the approximations that go into Eq. 5.1 – the Gaussian shape (hence the "1.6") and the ability to lump capacitance.

Equation 5.1 can also be rearranged by moving ΔV into the denominator under V_{gb}. When rearranged, Eq. 5.1 expresses the ground bounce as a fraction of ΔV. This is usually more useful since noise margin is also typically expressed as a fraction of ΔV and so the amount of ground bounce can more easily be compared to the noise margin. For example, if $V_{gb}/\Delta V$ is 0.15, then the ground bounce voltage is 15% of a full logic swing. By way of comparison, anything above about 10% of ΔV is usually considered marginal and may not work.

It must be understood that *this is a very rough approximation*; even simple SPICE simulations can show a different value of V_{gb}. This equation does give a useful approximation, however. If it indicates a ground bounce voltage near the noise margin of a digital input, the equation shows that you have a real problem that must be addressed.

Example 5.1

Consider a high-speed output on a digital chip. The output is driving 15 pF of capacitance, and the single ground connection is a bond wire inside an integrated-circuit package with 2.6 nH of inductance. (Remember we calculated this inductance in Chap. 4.) The rise and fall time of the output is 200 ps with a 1-V logic swing and the output has a resistance of 5 Ω.

(a) Estimate the total ground bounce seen on the ground connection.

First, we have to estimate the effective rise time of the output. The rise time of the gate is 200 ps but it is driving an output with an RC delay of $2.2 \times 5 \times 15 = 165$ ps. The effective rise time, then, is $\sqrt{200^2 + 165^2} = 259$ ps. So $V_{gb} = \frac{1.6*1*2600*15}{259^2} = 0.93$ V. (Notice the unit conversion of inductance.) This is far too high! The ground bounce is about 90% of a logic swing!

(b) Suggest ways to reduce the ground bounce and quantify the improvement.

From the equation, it is clear that a major source of ground bounce in this case is the inductance of the ground connection. Multiple grounds in parallel might reduce the ground bounce, but the inductance would have to drop by a factor of 10 to get V_{gb} below 10% of a logic swing. (In other words, we would have to use 10 wires in parallel.) This does not seem practical by itself.

Raising the rise time appears to be an effective method. The effective rise time must increase by a factor of 3 to reduce the V_{gb} by a factor of 9. So the effective rise time must increase to about 777 ps.

Using the rise time adder formula, $777 = \sqrt{200^2 + x^2}$. Solving for x, the RC delay must be 750 ps.

Using the risetime-RC formula, $2.2RC = 2.2 \times R \times 15pF = 750$ ps. So $R = 22.7$ Ω. Since the trace already has 5 Ω of resistance, a 17-Ω resistor can be added to the circuit. To give margin, and to move up to a standard resistor value, select a 22-Ω resistor. The RC delay increases to $2.2 \times 27 \times 15 = 891$ ps, the effective rise time increases to 913 ps, and so $V_{gb} = 0.079$ V, or about 8% of a full logic swing. This highlights a very common solution to ground bounce – add a series resistance in order to implement *rise time control*.

Results of Ground Bounce

By itself, the double-pulse would not be so bad. Every time the output changed, there would be a double-pulse on the ground connection on a falling edge and on the power connection on the rising edge. The next problem – and this turns out to be the problem that matters – is that the power and ground connection is shared by the rest of the chip that contains the driving gate. The driving gate is "polluting" the power and ground connection. (In fact, the power and ground connection of an output driver are often called "dirty power" and "dirty ground" for that reason.)

Two additional effects can then possibly occur.

First, consider a quiescent (unchanging or "static") output gate next to the one that is creating the falling edge. This is illustrated in Fig. 15.2.

There is a "changing gate" (a gate that has an output switching from 1 to 0) on the same chip as a "static gate" (a gate with a continuous 0 output). The changing gate has the equivalent circuit of a closing switch, and the static gate has the equivalent circuit of a short circuit to ground, a switch that has already closed. The gates are shown on the left, and the equivalent circuit is shown in the middle. Also note that three voltage-measuring points are marked, V_S, V_1, and V_2.

The changing gate works just like the gate up in Fig. 5.1, and the double-pulse waveform is created at point V_1. Recall the origin of the pulse – it is due to the inductance of the ground connection of the chip itself. So every point on the chip that is connected to this ground connection "sees" the double-pulse waveform.

Meanwhile, the changing gate produces a falling-edge waveform, indicated at point V_S in the figure. It is the same V_S as from the previous figure.

The static gate is driving a logic 0, which is a low-ohm connection to ground. More accurately, it is a low-ohm connection to the same point that now has a double-pulse waveform on it. Thus, the static gate actually carries the waveform to its output, and outputs the waveform to the gate that it is driving. The voltage of the static gate's output is measurable at point V_2, and the V_1 and V_2 waveforms are illustrated on the right.

This arrangement is odder than it looks. The changing gate produces the waveform shown at V_S, an ordinary falling edge. A neighboring gate, which should be producing a continuous 0-V (logic 0) output produces the waveform shown at V_2, a double-pulse. So the double-pulse waveform comes squirting out of the "quiet" logic-low neighboring output. This is the actual problem that ground bounce causes. In other words, the ground bounce itself is not the issue – it is what happens when the ground-bounce waveform comes out of quiescent outputs. This leads to an odd property of ground bounce – its appearance (double-pulse) is unmistakable, but it always shows up on a quiet, unchanging output.

This is usually how ground bounce is diagnosed – a quiet logic-low signal, sitting at 0 V, and suddenly belches out a double-pulse signal. In fact, I have seen it myself.

Engineer's Notebook: The Ground Bounce Pulse

We had designed our set-top box around a new type of plug-in card. The card had a 68-pin connector and a pretty high amount of power dissipation. We were confident, however, because the data bus ran at a fairly slow speed and we always overdesigned our power supplies. We plugged in the card and it soon became clear it did not work reliably. It was pretty clear that they (the card design team) thought that the problem was our set-top; after all, they had tested the card themselves.

So I went into the lab and started testing the card myself. I was just probing around some signals and then all of a sudden there it was: A double pulse waveform on a quiet signal output. The problem was immediately apparent: The card did not have adequate protection against ground bounce. The chip on the card was starving itself of ground current when the outputs switched. To be certain, I probed around and found out which outputs were switching at the moment of the double-pulse, which confirmed the diagnosis. (In other words, I recreated Fig. 5.2, with V_S and V_2 on my oscilloscope. To be clear, it looked almost exactly like Fig. 5.2!)

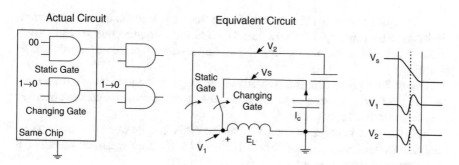

Fig. 5.2 The effect of ground bounce on a quiescent output

I was able to invoke the ground bounce recipe and got it working. The trick was more capacitance on the power and ground connections going up to the 68-pin connector, and series resistors on the data lines to get some rise-time control.

I could have blamed my co-workers who designed the card, because the ground bounce was coming out of their card, but it would have taken them a lot longer to fix, since their card was already in production. Besides, by adding the resistors to my set-top, it was ready in case other cards came along later with a worse ground-bounce problem.

Why is the double-pulse so bad?

First, recall that it is found on a "quiet" output, one that is not switching. If the pulse is high enough, it can rise above the noise margin of the logic input and look like a false logic transition. In other words, if ground bounce pulse coming out the quiet output is big enough, it might look like a fast digital pulse to the input that it drives. Since the signal is outside the noise margin, the input might be sensed as a pulse and it might not. Worse yet, it might manifest as an intermittent failure or as a failure after a change in part vendor. It might work well in the lab and turn into a huge headache once the product starts shipping, or once the production line changes from one brand of chip to another.

Second, if the ground bounce waveform becomes really big, the entire chip can "brown out." The voltage difference between power and ground is too small, and the chip does not work correctly. Memory can stop remembering, registers don't register, and so on. The chip stops working correctly.

Engineer's Notebook: The Self-Resetting Board

Once I was building up a board by hand so that students could connect and program an FPGA board. This was back before FPGA boards were available to mail-order, so we had to build our own. I found a truly ancient PC, so old that it had an old-fashioned ISA bus running at a whopping 14 MHz. I built up a board by hand to plug into the ISA slot in the PC, and I selected 74F family logic because I thought 74F was a very slow TTL logic family.

When we tested the board, it reset every time we used it. It did not take long to find the problem. 74F is not a slow TTL family; it is an extremely fast logic family.

The 74F part that was driving the PC bus had no decoupling and maximum-inductance hand wiring. (The power and ground were supplied over discrete hand-made wires that made big swirling loops. An inductance of 100nH would not be surprising.)

One of the "quiet" signals driven by the 74F part was the reset line. Sure enough, when one of the 74F part's output switched, the ground bounce on the reset line was large enough to trigger a reset.

This example really drove home to me the difference between fast clock speed and fast rise/fall times. Even at a glacially slow 14 MHz clock speed, the rise and fall times were so fast that ground bounce reset the board.

Ground bounce goes by several names. It is sometimes called *switching noise* or *delta-I noise*. It is made worse if several outputs switch at the same time, all in the same direction. In that case, *all* of the switching outputs create an inductive spike on the ground connection at the same time, and the final waveform is the sum of all of the spikes.

Does this situation occur in real life? The answer is unfortunately yes. Most digital systems are synchronous, with output transitions synchronized to clock edges. The clock rises, the D flip-flops latch new values, and all outputs switch at the same time. The number of *simultaneous switching outputs* is often used to estimate how bad ground bounce can become, and sometimes chip designers purposefully add clock skew (add delays to some of the output clocks) so that they don't all switch at the same time. Another, more common approach is to add separate power and ground connections for output drivers so that the chip's internal logic does not use the same power and ground. A typical rule of thumb is one power and ground connection for every six outputs. In fact, it is because of ground bounce that most chips have dozens of power and ground connections.

Minimizing Ground Bounce

So what can be done on the circuit board to reduce ground bounce? It always helps to go back and look at the formula. Ground bounce is linearly proportional to the inductance and capacitance and inversely proportional to the square of the rise time. Lowering the inductance usually means designing a better power supply, and so it is always good practice to minimize inductance.

A classic way to reduce the apparent inductance is to attach additional capacitance between the power and ground rail near the driving gate. When the power starts to sag or the ground voltage starts to rise (that is, when the difference between power and ground drops), the capacitor will start to release some of its charge, and the driving gate can "borrow" charge from the capacitor until the inductive power supply can catch up. Such capacitors near the chip are called *bypass capacitors*. These are often seen in pairs of large and small capacitors (such as 1 microfarad in parallel with 0.01 microfarad). The large capacitor is able to supply a lot of current (the instantaneous current demand can be hundreds of amps for a few nanoseconds) and the small capacitor is able to supply current up to very high frequencies (because the smaller size has less inductance).

Once inductance has been minimized, there may or may not be a way to reduce the capacitance seen by the signal on the driving gate. For example, if the driving gate is connected to an output that is far away, unless the board's layout can be changed, there is no way to reduce the capacitance substantially.

Finally, rise time is extremely important. A 10% increase in rise time results in a 20% reduction in ground bounce, for example. Many modern chips incorporate *rise time control* on the output pads for that reason. If a chip does not have rise time control, adding series resistors to outputs can RC-filter the outputs and make them slower, as was seen in Example 5.1. (Think of the series resistor as changing the fire hose to a soda straw; the chip sources or sinks substantially less current.)

One of the other advantages of series resistors is that they give you a way to change the circuit later without having to modify the circuit-board design. A new resistor value can be selected, which is much faster than having to lay a board out again.

To summarize, the two classic solutions to combat ground bounce are to add bypass capacitors to the board near chips with fast outputs and, if possible, add series resistors to fast outputs to slow them down.

Sometimes, however, the signals are designed according to some standard, like Serial ATA or PCI Express, and series resistors violate the standard. The good news is that chips that implement a published standard usually implement rise time control, and the standard itself usually calls out ranges of valid rise and fall times.

What Is Ringing?

Consider a logic gate driving an RC load through an inductor. In this case, the inductor is the inductance of the *signal trace*, the resistor is the sum of the gate's internal resistance and the resistance of the signal trace, and the capacitance is the sum of the signal trace's capacitance and the capacitance of the input that the logic gate drives. The three lumped elements form a series-RLC circuit. Because the circuit has both inductance and capacitance, it can exhibit a variety of behaviors.

First, we can define Q (short for *quality factor*) that defines how well the circuit oscillates. Q is a figure of merit for radio circuits in which oscillation is desirable, and so higher Q means the circuit oscillates better. (The "official" definition of Q is the ratio of energy stored to energy dissipated per cycle of oscillation.) In our digital world, high Q is undesirable of course, but sometimes occurs. For a series RLC circuit,

Fig. 5.3 Schematic of a
"typical" ringing circuit and
its SPICE simulation

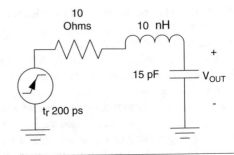

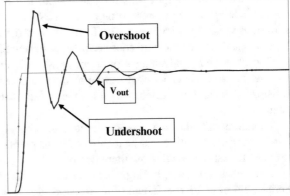

$$Q = \frac{\sqrt{L/C}}{R}. \tag{5.2}$$

If Q is greater than ½, the circuit is underdamped and can oscillate.

Second, at a specific frequency, the impedance of the inductor and capacitor cancel out. Solving for $\omega_0 L - 1/\omega_0 C = 0$, we find that $\omega_0 = 1/\sqrt{LC} = 2\pi f_{res}$ or

$$f_{res} = 1/2\pi\sqrt{LC}, \tag{5.3}$$

where f_{res} is the *resonant frequency* of the circuit. At that frequency, the series impedance of the circuit is minimized and is exactly equal to R. The circuit has the lowest impedance, and therefore lowest loss, at f_{res}. (We saw this once before, in Chap. 2 when discussing real capacitors.)

These are the two items that must be present for ringing to occur. First, the circuit must have a high value of Q. Second, the circuit must be "illuminated" (that is, driven by an outside source of power) at the resonant frequency. The high Q value is the bell and signal energy at f_{res} is the clapper that rings the bell. If a circuit has both items, it will ring like a bell. Figure 5.3 shows a typical ringing circuit and a simulated ringing waveform.

In Fig. 5.3, the voltage at the voltage source is the thin line and the voltage at "V_{out}" is the thick line, showing the ringing.

Results of Ringing

The result of ringing is the damped-sinusoidal waveform at the capacitor, which represents the circuit input. That leads directly to two important issues. First, the circuit input "sees" a bouncing input signal, and may see extra logic transitions if the bounce is high enough. Second, the circuit emits signal energy at f_{res} which may be important if the signal trace acts like an antenna. It can cause interference with other systems and may cause the system to fail radiated emissions testing. (We will discuss radiated emissions later.)

One important point about f_{res} is that it has no correlation to any clock frequency in the system. It appears as radiated emissions at a "random" frequency, which can make it difficult to track down.

In the underdamped region of interest, the output voltage is a damped sinusoidal waveform. Thus, the voltage is the product of an exponential term, with a negative exponent, and a sinusoidal term. To be specific, and to make a long story short, it has the form $\Delta V e^{-\alpha t} \sin(\omega_d t + \varphi)$ where $\alpha = R/2\,L$ and $\omega_d = \alpha\sqrt{4Q^2 - 1}$.

In the presence of a step input (with non-zero rise time), the sinusoidal function reaches a local maximum at $\omega_d t = \pi$ and so the waveform reaches a maximum at

$$\frac{V_{overshoot}}{\Delta V} = e^{-\alpha t} \sin(\pi) = e^{-\frac{\omega_d t}{\sqrt{4Q^2-1}}} = e^{-\frac{\pi}{\sqrt{4Q^2-1}}} \tag{5.4}$$

The sinusoid continues oscillating and then reaches a minimum at

$$\frac{V_{undershoot}}{\Delta V} = e^{-\alpha t} \sin(2\pi) = e^{-\frac{2\pi}{\sqrt{4Q^2-1}}} = \left(\frac{V_{overshoot}}{\Delta V}\right)^2 \tag{5.5}$$

The "undershoot" is actually more serious; it is the point where the ringing has caused a maximum reversal of the logic transition. For example, on a low-to-high transition, it is the point where the voltage "bottoms out" and almost looks like a logic 0. As was the case with ground bounce, if the undershoot is a significant fraction of the signal swing, the logic input might consider it an extra, unwanted logic transition.

Consider the following example.

Example 5.2

A student is probing an old hand-wired circuit board. The circuit has an inductance of 100 nH and is driving a load of 30 pF. The source impedance is 10 Ω and the source has a fall time of 2 ns. (Since it is TTL, the fall time is much faster than the rise time.) The board works at a clock frequency of 10 MHz.

(a) What is the Q and f_{res} of the circuit?

$$Q = \frac{\sqrt{100n/30p}}{10} = 5.77 \text{ and } f_{res} = \frac{1}{2\pi\sqrt{100n \times 30p}} = 91.9 \text{ MHz}$$

Note the use of nano and pico in calculating the numerator of Q – the units are very important to keep straight.

(b) Will the circuit ring?

Q is clearly high enough (it is much greater than 1), but is the resonance illuminated at the resonant frequency?

One might be tempted to look at the clock frequency (10 MHz) and conclude that it was much lower than the resonant frequency (92 MHz), and call it a day. But this is a mistake.

Consider instead the circuit fall time of 2 ns. From that we can compute $f_{knee} = 1/4$ ns $= 250$ MHz. The digital signal has significant signal energy out to 250 MHz (in fact, significant energy out to 25 times that of the clock frequency). The knee frequency is well past the resonant frequency, and so, yes, the resonance is illuminated and the circuit will ring. (Another example of why knee frequency is so important!)

(c) What will the signal look like on an oscilloscope?

It will look like a falling edge followed by a damped 91.9 MHz sinusoid.

(d) What can be done to reduce ringing?

One way to eliminate ringing is to band-limit (that is, low-pass filter) the digital signal below the ringing frequency. To get the knee frequency below 91 MHz, the fall time must be above 1/182 MHz $= 5.4$ ns. This requires an RC filter with a rise time of $\sqrt{5.4^2 - 2.0^2} = 5$ ns. Assuming the capacitive load stays the same, this means the series resistance must increase to a total of $R = 5 \text{ ns}/2.2 \times 30 \text{ pF} = 75.7 \ \Omega$, so a 66-$\Omega$ resistor must be added to the circuit. Since the next-highest standard resistor value is 68 Ω, select that.

An alternative is to lower Q below $\frac{1}{2}$. Rearranging Eq. 5.2, we have $R = \frac{\sqrt{L/C}}{Q} = \frac{\sqrt{100n/30p}}{0.5} = 115 \ \Omega$ or adding a 105-Ω resistor to the circuit.

In either case, the solution is to add series resistance to the circuit to slow it down. That is a common theme in solving lumped signal-integrity issues.

Minimizing Ringing

As we saw in the example, one way to minimize ringing is to raise R. An alternative, based on Eq. 5.2, is to lower L. Technically speaking, raising C would also help but is usually undesirable because it slows down the signal.

So Where Is Ringing Seen – And Not Seen – Today?

As indicated in the introduction to the chapter, ringing is rarely seen on signals because rise times have dropped to the point that most signals are distributed. In the world of distributed signals, the signal trace acts like a resistor instead of an inductor and capacitor (more on that later) and so L is effectively close to 0.

There is one place where ringing is seen very often, and that is when probing signals incorrectly (or in a hurry). Oscilloscope probes have long ground wires, which add a lot of inductances, and the scope has a low input capacitance and resistance on purpose. Additionally, the scope's input bandwidth can be so low as to make the circuit seem lumped. (More accurately, the distributed behavior of the circuit is low-pass-filtered out.) The result is that improperly placed scope probes can create the appearance of ringing where there isn't any. If you see ringing on an oscilloscope and are worried about it, switch over to a high-frequency probing method (such as removing the witch's hat on the scope probe and connecting to a ground close to the signal) and see if the ringing goes away. If the ringing drops substantially, it is an artifact of the scope. You can go back to the quick-and-dirty probing method and do not need to worry about ringing.

There is one physical phenomenon commonly confused with ringing, reflections on a transmission line. We will discuss transmission lines later, but for now it is enough to say that if the reflection coefficients at the source and load are of opposite sign and their product has a magnitude close to 1, the load (the logic input) waveform will bounce up and down just like ringing is occurring. However, this bouncing waveform is NOT caused by Q or resonance; it is caused by the impedance mismatch. It is very important to understand this distinction. Many textbooks in the field get this completely wrong, and call the mismatched-impedance behavior "ringing" but IT IS NOT RINGING. It has a completely different underlying physical cause and therefore is diagnosed and fixed through a completely different process (one involving, ironically, series resistors). This is the "other reason" why we studied ringing – so that when we study reflections, you can see that it is NOT ringing.

Appendix

1. Estimate the ground bounce for the following situation.

   - Write down the last three digits of your Student ID number:

     _____ _____ _____

   - Replace any instances of '0' with the last non-zero digit in your ID number:

   Digit R _____ L _____ C _____

   - Voltage source rises from 0 to 1 V with a 10–90% rise time (T_{source}) of 150 ps
   - Series resistance of 5*R Ω
   - Series inductance of 5*L nH
   - Circuit load is 5*C pF.

   (a) Estimate the rise time of the circuit.

       (i) First, compute the 10–90% rise time of the RC constant: $T_{rc} = 2.2RC$
       (ii) To find the overall rise time, combine the rise time of the source with the

       rise time of the RC constant as follows: $T_{rise} = \sqrt{T_{rc}^2 + T_{source}^2}$.

   (b) Estimate the amount of ground bounce using the formula $V_{gb} = \frac{1.52LC\Delta V}{T_{rise}^2}$.

2. The ground bounce waveform is a "double pulse." It is the second derivative of which waveform?

3. Consider the ground bounce formula $V_{gb} = \frac{1.52LC\Delta V}{T_{rise}^2}$. What four things can be done to reduce ground bounce?

4. Consider a circuit with an output resistance of 10 Ω and a ground-trace inductance of 10 nH driving a load of 25 pF.

   (a) What is the Q of the circuit?
   (b) What is the expected ringing frequency?
   (c) If the circuit rings, what is the expected overshoot?
   (d) How can you lower the Q of the circuit?

5. A circuit has an output resistance of 5 Ω, a ground-trace inductance of 10 nH, and is driving a load of 10 pF.

   (a) What is the Q of the circuit?
   (b) What is the expected ringing frequency?
   (c) Will the circuit ring if the output has a rise time of 500 ps?
   (d) What is the minimum rise time that keeps the knee frequency of the signal below the ringing frequency?

6. Consider a circuit that has 10 Ω source resistance, a 10 pF load, and 25 nH of inductance on the ground connection. The signal has a 1 ns risetime.

   (a) Estimate the ground bounce as a fraction of the full logic swing. Use the formula $\frac{V_{gb}}{\Delta V} \approx \frac{1.5LC}{T_{10-90}^2}$

(b) Estimate the Q of the circuit.

(c) Calculate the ringing frequency and knee frequency of the circuit.

(d) Is the knee frequency above or below the ringing frequency? Will the circuit ring? Why or why not?

7. Estimate the ground bounce for the following situation.

- Voltage source falls from 1 to 0 V with a 10–90% rise (fall) time (T_{source}) of 333 ps
- Series resistance of 5 Ω
- Series inductance of 20 nH
- Circuit load is 30 pF.

(a) Estimate the rise time of the circuit.

(i) First, compute the 10–90% rise time of the RC constant: $T_{rc} = 2.2RC$ $2.2 \times 5 \times 30 = 330$ ps

(ii) To find the overall rise time, combine the rise time of the source with the rise time of the RC constant as follows: $T_{rise} = \sqrt{T_{rc}^2 + T_{source}^2}$ sqrt $(333^2 + 330^2) = 469$ ps

(b) Estimate the amount of ground bounce using the formula $V_{gb} = \frac{1.6LC\Delta V}{T_{rise}^2}$.

$\Delta V = 1$. $V_{gb} = 1.6 \times 20{,}000p \times 30p \times 1/(469p)^2 = 4.36$ V.

(c) Is the ground bounce serious? That is, is $V_{gb}/\Delta V$ greater than 10%? It's about 400%! Yes, extremely serious.

8. The ground bounce waveform is a "double pulse." It is the second derivative of which waveform? Rising/falling edge

9. Consider the ground bounce formula $V_{gb} = \frac{1.52LC\Delta V}{T_{rise}^2}$. What four things can be done to reduce ground bounce? Lower L, Lower C, Lower ΔV, Increase T_{rise}

10. Consider a circuit with an output resistance of 5 Ω and a ground-trace inductance of 20 nH driving a load of 30 pF (same as question 1).

(a) What is the Q of the circuit? $Q =$ sqrt(20,000/30)/5 = 5.16

(b) What is the expected ringing frequency? $f_{res} = 1/(2 \times 3.14 \times$ sqrt $(20000p \times 30p)) = 205$ MHz

(c) Will the circuit ring if the output has a rise time of 333 ps? $f_{knee} = 1/0.666$ ns = 1.5 GHz. Yes.

(d) What is the minimum rise time that keeps the knee frequency of the signal below the ringing frequency? $1/(2 \times 205$ MHz$) = 2.4$ ns

(e) If the circuit rings, what is the expected overshoot?

From the book: $\frac{V_{overshoot}}{\Delta V} = e^{-at} \sin(\pi) = e^{-\frac{\omega_d t}{\sqrt{4Q^2-1}}} = e^{-\frac{\pi}{\sqrt{4Q^2-1}}}$

$4Q^2 - 1 = 105.5$. $(-3.14/$sqrt$(105.5)) = 0.305$ $e^{-0.305} = 0.736$.

(f) How can you lower the Q of the circuit? Lower L, Raise C, Raise R.

Chapter 6
Distributed Analysis: Transmission Lines and Z_0

Background and Objectives

This chapter introduces distributed analysis, the art of analyzing circuits and structures that are so physically large that the voltage on wires is not necessarily the same at each point. The fact that the structure is so large means that parts of the circuit are "unaware" of what other parts are doing, and the results are, to say the least, counterintuitive. Wires turn into resistors and excesses in current cause voltage waves to crash back and forth. When you finish this chapter, you should be able to:

- Explain what a transmission line is
- Explain the derivation of characteristic impedance (Z_0) and the physical consequences of its existence
- Understand the approximate model of a transmission line

Where Can You See a Transmission Line?

Maxwell's equations predicted a spooky world where energy could radiate outward from a point, move at a certain, fixed speed, and cause voltages to appear some distance away. Heinrich Hertz built one of the first systems to exploit this strangeness, creating the radio.

Albert Einstein used Maxwell's notion of the speed of light to construct a model of the universe that is relativistic. In effect, Einstein's theory outlines a universe where nothing is actually "simultaneous." A logic gate driving an output cannot "see" what happens at the other end instantaneously. For small wires or for slowly changing voltages, the net effect is nearly instantaneous, and this is the familiar "lumped" model of circuits. We are entering the domain of electrical engineering where this is no longer true – changes to circuits do not occur instantaneously. Like the old saw "the left hand does not know what the right hand is doing," a distributed

© Springer Nature Switzerland AG 2022
S. H. Russ, *Signal Integrity*, https://doi.org/10.1007/978-3-030-86927-4_6

circuit is one where the source of a changing voltage (or current) does not "know" what is happening at the destination. So now there is not only "spooky action at a distance," but there is also the fact that the results of the action at the distance are not known until later.

Getting back to radio... A radio uses a transmission line. In the case of a radio, the transmission line is the air between the two antennas (transmitting and receiving). A cable or DSL modem uses a transmission line to send and receive data, with the transmission line being, respectively, the coaxial cable and twisted-pair telephone line.

What do these examples have in common? First, the signal energy is propagating in a dielectric, an electrical insulator. Second, there is a relatively low loss of signal in the dielectric. This means that the signal tends to arrive intact at the other end, which makes it much easier to receive.

There are some differences. For example, the radio signal gets weaker farther from the antenna, but only because the energy is spreading out in space (i.e., the dielectric is getting larger as the signal moves away from the antenna). In the case of twisted-pair or coaxial cable, the signal energy is confined to the structure and so gets weaker at a much slower rate.

So how do transmission lines work? To understand that, consider a small piece of coaxial cable...

Transmission Line: The View from the Inside

Let's start with 1 m of ordinary RG-11 coaxial cable, commonly used for cable television or to carry satellite signals from the dish to the satellite set-top box. The term "coaxial" refers to the fact that it has a small center conductor surrounded by a cylinder of insulating material, which is in turn surrounded by a cylindrically shaped conductor. The two conductors are coaxial (get it?) and so they make a capacitor. If you measured the capacitance of 1 m of RG-11, it turns out to be 53.1 pF. If you measure its inductance, it is 298.7 nH. (This information is posted in datasheets for the cable.)

It is worth noting that if you double the length to two meters, both the capacitance and inductance go up by a factor of two. So we can say that RG-11 has 53.1 pF/m of capacitance and 298.7 nH/m of inductance. Stated symbolically, $C_L = 53.1$ pF/m and $L_L = 298.7$ nH/m.

What happens when one launches a very fast digital signal, say a 50 ps rising edge, down a long piece of RG-11 cable?

First, before the rising edge is launched, the voltage of the center conductor is less than the voltage of the outer conductor. Because of capacitance, the difference in voltage results in an E field between them.

Second, the rising edge "flips" the E field because, after the rising edge passes, the voltage of the center conductor is now higher. To be specific, and this is very important, *it is the reversal of the E field that travels at nearly the speed of light.*

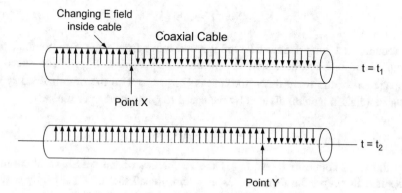

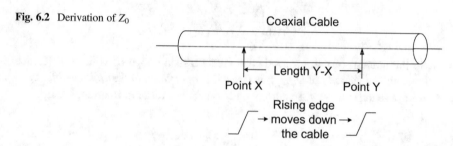

Fig. 6.1 Motion of a rising edge down a transmission line. Note that the edge passes point X at time t_1 and point Y at time t_2

Fig. 6.2 Derivation of Z_0

(To review, if the permittivity is higher than that of free space, light moves more slowly, and the E change in the E field moves at a slower speed.) It is exactly and specifically *not* the electrons that are moving that fast.

This is illustrated in Fig. 6.1.

The rising edge (more accurately, the changing E field) has to move from point X on the cable to point Y on the cable. The E field can only change if real electrical charge is applied to each region of the cable. To be clear, the E field can only change if the electrical charges on that region of the capacitor (the one formed between the two conductors at the point of the rising edge) are reversed, and the charge can only be reversed by applying electric current to redistribute the charges. Charge is applied at a certain rate, which of course is known as current. Since the changing E field is constantly moving down the cable, more current has to be put into the cable in order for the changing E field to continue down the line. One can then calculate the amount of current needed to enable the rising edge to race down the line. (Also see Fig. 6.2 and follow along.)

First, how much electric charge is needed to charge the piece of cable between point X and point Y? That is a simple calculation: charge is the product of the capacitance and the voltage. The capacitance is the length between points X and Y multiplied by the capacitance per unit length. The voltage is the voltage swing from a logic 0 to a logic 1, which is denoted "V".

$$Q = (Y - X)C_L V \tag{6.1}$$

Second, how long does it take the wave to move from point X to point Y? It is simple once one recalls Maxwell's formula for velocity, $1/\sqrt{\mu\varepsilon}$. The speed of propagation of the wave down the line is $1/\sqrt{C_L L_L}$ and so the time interval is the distance (e.g., in meters) divided by the speed (e.g., in meters per second).

$$t = (Y - X)\sqrt{C_L L_L} \tag{6.2}$$

(By the way, notice that $\sqrt{C_L L_L}$ is the *reciprocal* of the speed, so if speed is measured in cm/ps, the reciprocal is in ps/cm. Recall that this is a handy way to measure speed because it lets one easily calculate the delay given the length.)

So what is the current?

$$I = \frac{Q}{t} = \frac{(Y - X)C_L V}{(Y - X)\sqrt{C_L L_L}} = V\sqrt{\frac{C_L}{L_L}} \tag{6.3}$$

So the amount of current needed to create a rising edge of voltage difference V traveling at a velocity of $1/\sqrt{C_L L_L}$ is *linearly proportional to the voltage difference*. A linear relationship between current and voltage is a resistor. Rearranging Eq. 6.3, we get

$$V = I\sqrt{\frac{L_L}{C_L}} \equiv IZ_0 \tag{6.4}$$

and so

$$\sqrt{\frac{L_L}{C_L}} \equiv Z_0 \tag{6.5}$$

Again, not to put too fine a point on it, Z_0 is *the ratio of the desired voltage swing to the current needed to create that swing on a distributed line with capacitance per length C_L and inductance per length L_L.* In other words, it is the ratio of voltage swing to current on a distributed line.

The last part (distributed line) is important and easily overlooked. If you look back at Chap. 1, the distributed/lumped boundary involved the length of a rising edge relative to the length of the wire carrying the rising edge. The significance here is that Z_0 is only relevant while the rising edge is racing down the line. When viewed over a longer period of time, the wire turns back into a lumped wire with zero resistance.

Z_0 is a very easily misunderstood concept, so notice the ingredients. First, as just explained, the signal must be operating in a distributed regime (meaning that the length of the connection must be greater than the length of the signal transition).

Second, the signal must propagate in a medium with the capacitance per length and inductance per length as designated. Third, the capacitance and inductance dictate the speed at which the signal travels, which in turn dictates the rate at which electric charge must be pumped into the line. In other words, it dictates the current (rate of electric charge) that is needed. Fourth, electric charge is being pumped into the line in order to charge the capacitance per unit length, and it must continue to be pumped in if the signal is to continue down the line. The rate of charge is equivalent to current. Fifth, notice how units of distance cancel out.

To the gate that is driving a fast-rising edge down the line, the long piece of coaxial cable *looks like a real resistor of value Z_0 to ground*. The "wire" has turned into a "resistor". The resistance is purely real – not imaginary – which is highly counter-intuitive. Even though the wave propagation is due to inductance and capacitance, the current that is being sourced is charging a physical capacitance and so involves the transfer of actual electric charge. It is a real impedance, and *Z_0 is purely real*.

Example 6.1

What is the characteristic impedance of RG-11?

$Z_0 = \sqrt{\frac{298.7n}{53.1p}} = \sqrt{\frac{298700}{53.1}} = \sqrt{5625} = 75\,\Omega$. This is why RG-11 is referred to as "75-Ω cable". (Notice the units conversion needed on the inductance in order to perform division.)

As if this was not strange enough, consider this: what, exactly, is moving down the line and where is it located?

The "what exactly": According to Maxwell's equations, the "thing" that is moving down the line at speed $1/\sqrt{C_L L_L}$ is an electromagnetic wave, a changing E field. It is specifically NOT the electrons themselves – they actually move at a much slower pace. Just like waves can ripple across an ocean with the water remaining basically motionless, electromagnetic waves can move much, much faster than electrons.

The "where": Remember that waves in a transmission line move *in the dielectric*. (If you are hazy on this, remember how your cell phone works!) In our RG-11 example, the rising edge (again, the changing E field) is NOT located in the center conductor; it is located in the dielectric between the two conductors.

On a circuit board, this change becomes crucial to the thought process that accompanies good design. The rising edge is *located in the fiberglass between the signal and the ground plane*. Any imperfection in the fiberglass *or* the ground plane will distort the signal. At high frequencies (more accurately, at a wire length greater than $\ell/6$), the signal is no longer in the wire – it is in the *combination* of the wire, the insulator, and the ground plane. And so any perturbation to this tangled web of signal propagation (any change to the geometry of the wire, insulator, or ground plane) results in signal distortion.

There is one aspect of this newly interconnected world that must be emphasized forcefully. The signal and its return current are carried in the dielectric, and so the signal trace and ground plane are now *equally responsible* for carrying the signal.

"Ground" (i.e., the ground plane) is no longer a DC signal reference or part of the power infrastructure; it is the second conductor of a matched pair that is needed to carry the signal.

Once logic gate has launched a fast-rising edge down a long piece of RG-11 coaxial cable, how long does it take to reach the end? One can look back at Eq. 6.2. Defining T_0 as the one-way transmission time and l as the length of the cable, we have

$$T_0 = l\sqrt{C_L L_L} \tag{6.6}$$

Example 6.2

What is one-way transmission time down 2.5 meters of RG-11?

$T_0 = 2.5\sqrt{53.1p \times 298.7n} = 2.5 \times 3.98n = 9.96$ ns. The speed of light in a vacuum is 33.3 ps/cm so light traverses 2.5 m of vacuum in $33.3 \times 250 = 8325$ ps= 8.325 ns. So electromagnetic waves travel about 20% slower in RG-11 than in free space.

Estimating Z_0 in Common Situations

Z_0 is a function of a transmission line's inductance and capacitance per unit length. To calculate Z_0 precisely for a transmission line usually requires sophisticated electromagnetic simulation, but fortunately there are some formulas that estimate Z_0 for commonly encountered circuit-board structures. These are approximations (actually, they are industry-standard approximations), but fortunately "approximately-correct" termination results in "nearly-perfect" signal transmission. It is a situation where being off by a few percent will usually work correctly.

The formulas were published by IPC, a standards-making body for the printed circuit board and surface-mount industry, and unfortunately are based on English units. First, for microstrip (which is a conductor on the top layer of a circuit board over a ground plane separated by a dielectric):

$$Z_0 = \frac{87}{\sqrt{\varepsilon_r + 1.41}} \ln\left(\frac{5.98h}{0.8w + t}\right) \tag{6.7}$$

$$T_p = 85\sqrt{0.475\varepsilon_r + 0.67} \tag{6.8}$$

$Z_0 =$ Characteristic Impedance (ohms)
$T_p =$ Delay in ps/in
$w =$ conductor width (mils)
$t =$ conductor thickness (mils)
$h =$ spacing from conductor to plane (mils)

Second, for stripline, which is a conductor centered vertically between two planes and completely surrounded by a single, uniform dielectric:

$$Z_0 = \frac{60}{\sqrt{\varepsilon_r}} \ln \left(\frac{1.9b}{0.8w + t} \right) \qquad (6.9)$$

$$T_p = 85\sqrt{\varepsilon_r} \qquad (6.10)$$

Z_0 = Characteristic Impedance (ohms)
T_p = Delay in ps/in
w = conductor width (mils)
t = conductor thickness (mils)
b = spacing from plane to plane (mils)

Example 6.3
A line is laid out in microstrip topology using FR-4 fiberglass ($\varepsilon_r = 4.2$). The line is 5 mils wide and lies 5 mils above the ground plane. The trace is constructed from "1-ounce" copper which is 35 μm (1.4 mils) thick. What is the Z_0 and T_p of the trace?

$$Z_0 = \frac{87}{\sqrt{4.2 + 1.41}} \ln \left(\frac{5.98 \times 5}{0.8 \times 5 + 1.4} \right) = 36.7 \ln (5.53) = 62.8 \ \Omega$$

$$T_p = 85\sqrt{0.475 \times 4.2 + 0.67} = 85\sqrt{2.665} = 138.8 \ \text{ps/in}$$

5-mil-wide traces and 1-ounce copper are typical values for most modern circuit boards. The designer then chooses the thickness of the insulating layer between the signal layer and the ground layer to get the desired Z_0.

Approximate Model of a Transmission Line

There is another way to think about transmission lines, based on a model of how they work. Consider a pair of conductors in close proximity that travel over a long distance. There is a mutual capacitance between the conductors because they can be at different voltages and there is insulating material between them, the conductors have mutual inductance because they are physically separated and their fluxes do not completely cancel, and the conductors have series resistance. Also, because the insulating material is not a perfect insulator, there is a conductance between the two conductors. (This is called "dielectric loss" – more on this later.) The model is shown in Fig. 6.3.

To model a transmission line, it is divided into a very large number of tiny segments, and each segment is assigned some mutual capacitance, mutual inductance, resistance, and conductance. Using this model, all the behaviors (such as Z_0)

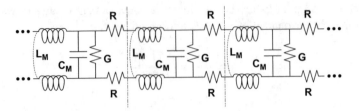

Fig. 6.3 Lumped model of a transmission line

emerge. This model does not help us understand the physical significance of Z_0 but it helps programmers simulate transmission lines.

By the way, this model leads to "the telegrapher's equation" because telegraph wires were among the first transmission lines ever developed.

It is helpful to describe the four terms (C, L, R, and G) as an amount per unit length. When this is done, it is common to add a subscript L to indicate that the unit is per length. Capacitance C, for example, becomes C_L and has units like pF/cm.

This model will come in handy in the chapter on dielectric loss and lossy transmission lines.

Appendix

1. Consider a transmission line.

   (a) What structures or features cause impedance discontinuities and therefore signal reflections? (Name three examples.)
   (b) Where is the signal located as it propagates?
   (c) What factors determine a transmission line's characteristic impedance?

2. A vendor has submitted a cable for you to evaluate for use in a high-speed signal environment. When measuring 10 inches of cable, you find that it has 64 nH of inductance and 11.4 pF of capacitance. What is the cable's characteristic impedance?

3. Consider a transmission line with impedance $Z_0 = 50\ \Omega$.

   (a) What is the phase angle between the voltage and the current? Stated differently, is the 50-Ω impedance seen at the source real, complex, or imaginary?
   (b) If the trace was moved farther from the ground plane, would you expect Z_0 to go up or down?

4. Consider a 5-mil-wide circuit trace located between two ground planes that are 10 mils apart. (That is, the two planes are 10 mils apart and the circuit trace is centered between them.) The dielectric is FR-4 ($\varepsilon_r = 4.2$). The trace is routed stripline, completely surrounded by the FR-4 and the two planes. The trace is 9 inches long and is carrying a signal with a 200 ps rise time.

(a) What is the knee frequency of the signal?
(b) What is the speed of propagation of a signal on the trace, in ps/in?
(c) How long must the trace be in order to be considered distributed?
(d) What is the Z_0 of the trace?
(e) What are the T_p and T_0 of the trace?

5. Consider a transmission line with impedance $Z_0 = 50 \, \Omega$. The transmission line is a trace on a circuit board.

   (a) What is the phase angle between the voltage and the current? Stated differently, is the 50-Ω impedance seen at the source real, complex, or imaginary? Real
   (b) If the trace was moved farther from the ground plane, would you expect Z_0 to go up or down? Farther, more inductive, Z_0 goes up
   (c) If the trace was made narrower, would you expect Z_0 to go up or down? Narrower = more inductance and less capacitance. Z_0 goes up.

6. Consider a 5-mil-wide circuit trace located 10 mils over a ground plane. The dielectric is FR-4 ($\varepsilon_r = 4.2$). The trace is routed microstrip, on top of the FR-4 and ground plane. The trace is 25 cm long and is carrying a signal with a 200 ps rise time.

   (a) What is the knee frequency of the signal? 1/400 ps = 2.5 GHz
   (b) What is the Z_0 of the trace? (Use the formula from the slides.) I get 88.3 Ω...
   (c) What are the T_p and T_0 of the trace? T_p = 138.8 ps/in = 54.6 ps/cm. T_0 = 54.6 ps/cm × 25 cm = 1.366. ns.

Chapter 7
Transmission Lines: Reflections and Termination

Background and Objectives

This chapter introduces the concept of terminating a transmission line. Because of the "left hand does not know what the right hand is doing" property of transmission lines, changes in voltage can have unexpected effects at the source and load (and points in between) in the form of reflections. Managing these reflections requires a termination strategy. When you finish this chapter, you should be able to:

- Describe how reflections occur on improperly terminated lines
- Calculate and simulate simple transmission lines
- Calculate the correct impedance to terminate transmission lines
- Describe different termination strategies and their relative merits
- Explain how to diagnose and deal with common termination issues

The Left Hand and the Right Hand: Reflection and Transmission

Reviewing briefly, if a gate launches a very fast rising edge into RG-11 coaxial cable, the gate "sees" a resistance of 75 Ω to ground. The rising edge then takes 10 ns to get to the end of 2.5 m of cable.

What happens next is completely unexpected. There are two basic issues in play.

First, the 75-Ω impedance of the cable is in series with whatever resistance is at the source and at whatever resistance is at the other end. (We call the resistance at the other end the *load* or *load resistance*.)

Second, the source is isolated from the load by the relative slowness of the speed of light. It is as if the signal is being carried by carrier pigeons and takes days to reach the other end. The source does not "know" what is happening at the load, at least not at first. In our RG-11 example, it takes 10 ns for the edge to reach the load and then

© Springer Nature Switzerland AG 2022
S. H. Russ, *Signal Integrity*, https://doi.org/10.1007/978-3-030-86927-4_7

Fig. 7.1 Currents and
voltages used to derive the
reflection-coefficient
formula. Note that the
resistor of Z_0 to ground is
how the transmission line
appears to the load

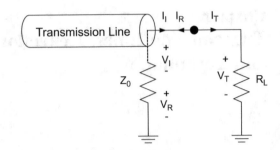

another 10 ns for news of what happened to get back to the source. Because we are
operating well above the critical length, the theory of relativity makes clear that there
is no way for the source to have instantaneous knowledge of the load. Again, this is
what makes the circuit distributed instead of lumped: the "left hand" (source) does
not know what the "right hand" (load) is doing.

So what happens when the wave hits the load? It depends on the load.

If the impedance of the load is greater than the impedance of the line (i.e., if
$Z_L > Z_0$) then the rising edge traveling down the line has too much current for the
desired voltage swing at the load. The excess current crashes (more accurately, it
reflects) back onto the line in the form of a positive reflection. The reflection is
positive because there is an excess coming back off of the load.

If the impedance of the load is less than the impedance of the line (i.e., if $Z_L < Z_0$)
then the rising edge traveling down the line does not have enough current for the
desired voltage swing at the load. The shortage of current appears in the form of a
negative reflection off of the load. One can imagine it as if the load is sending back to
the source for more current.

So how can we model this? Referring to Fig. 7.1, an *incident* current and voltage
emerge out of the line and strike the load. Part of the voltage and current are *reflected*
back onto the line and part is *transmitted* onto the load. We will denote the incident
wave with the subscript I, the reflected wave with R, and the transmitted wave with
T. Our goal is to calculate the ratio of the reflected voltage (V_R) to the incident
voltage (V_I).

The line has an impedance Z_0 and the load has an impedance R_L. We assume here
that the load is purely resistive, although the math works out exactly the same if it is
not. Note that we do not have to assume that Z_0 is purely real – it *is* purely real!

I_I is coming out of the line and I_R is going back onto the line, and so we know that
$I_I = V_I/Z_0$ and $I_R = V_R/Z_0$. Likewise, $I_T = V_T/R_L$ because it is flowing through the load
resistance. In other words, the resistances of the line and of the load constrain the
respective ratios of current to voltage.

The voltage at the end of the line must equal the voltage across the load, since the
end of the line and the load have the same node. We can apply lumped analysis here
because the load is lumped, and the coaxial cable's Z_0 has the illusion of being a
lumped resistance. In other words, this analysis is being applied at an instantaneous
moment in time so that we can use lumped analysis. We know that

$$V_I + V_R = V_L \tag{7.1}$$

because the voltage of the load equals the voltage at the line (they are the same node) and the voltage of the line equals the sum of the incident and reflected voltages.

The incident current must equal the transmitted and reflected current. So we know that $I_I = I_R + I_L$. (If you study everything to this point, the sign for V_I looks backward relative to I_I, but that is because the incident wave is a source of current and the reflection and transmission are sinks of current.)

Substituting the three formulas for current into the current equation, we have

$$V_I/Z_0 = V_R/Z_0 + V_T/R_L \tag{7.2}$$

and so

$$V_I - V_R = \frac{Z_0 V_T}{R_L} \tag{7.3}$$

If we subtract and add Eqs. 7.1 and 7.3, we get, respectively,

$$2V_R = V_T\left(1 - \frac{Z_0}{R_L}\right) \tag{7.4}$$

$$2V_I = V_T\left(1 + \frac{Z_0}{R_L}\right) \tag{7.5}$$

So dividing Eq. 7.4 by Eq. 7.5, and canceling out, we get

$$\frac{V_R}{V_I} = \frac{\left(1 - \frac{Z_0}{R_L}\right)}{\left(1 + \frac{Z_0}{R_L}\right)} = \frac{R_L - Z_0}{R_L + Z_0} \equiv \rho_L \tag{7.6}$$

So we define the ratio V_R/V_I as the *reflection coefficient at the load* and denote it ρ_L. It lets us calculate the reflected voltage as a function of the incident voltage.

To summarize the derivation, by expressing the incident, reflected, and transmitted waves and their respective impedances, we can "balance the books" and construct a model for what happens to the signal energy.

If $R_L > Z_0$, then $\rho_L > 0$ and there is a positive reflection (and as $R_L \to \infty, \rho_L \to 1$). Likewise, if $R_L < Z_0, \rho_L < 0$ and there is a negative reflection. If $R_L = Z_0, \rho_L = 0$ and there is no reflection. This is in line with the earlier statement that if $R_L > Z_0$, there is an excess of current and a positive reflection.

We can also define the ratio V_L/V_I as the *transmission coefficient at the load* and denote it T_L. Dividing Eq. 7.6 by V_I, we get

$$\frac{V_I + V_R}{V_I} = 1 + \rho_L = \frac{V_T}{V_I} = T_L \qquad (7.7)$$

In other words, $1 + \rho_L = T_L$. T_L lets us calculate the transmitted voltage (the voltage appearing at the load) as a function of incident voltage.

Let's review what happens. When the waveform hits the "other end" of the transmission line, part of the waveform is reflected and part is transmitted. The transmitted part becomes visible as a voltage across the load. The reflected part goes back onto the transmission line and goes back to the source.

Sometime later (T_0 later, to be exact), the reflected wave hits the *source*. The source has a similar reflection and transmission coefficient, and, once again, part of the energy is transmitted, becoming a voltage visible at the source, and part is reflected, heading back to the load again.

The result is a pinball game of signal energy, with the size of the pinball a function of the reflection coefficients at the source and load.

So let's work an example to see what happens. . . (Example 7.1)

Example 7.1

Consider a long 50-Ω transmission line on a circuit board. A fast-falling edge is being driven on the line by a source with a 10-Ω resistance. At the other end of the line, the receiving logic gate looks like a 5 MΩ resistance to ground. (The source and load resistances are typical for CMOS logic.) The voltage swing is 4 V, and both the source and load are quietly resting at +4 V when everything starts. Plot the source and load voltages in time units of T_0.

First we first need to know the source and load reflection and transmission coefficients. $\rho_L = (5\text{Meg}-50)/(5\text{Meg} + 50) = 1$. $T_L = 1 + \rho_L = 2$. $\rho_S = (10-50)/(10 + 50) = -2/3$. $T_S = 1 + \rho_S = +1/3$.

Then we need to consider what happens to the wave when it is launched. The source is a square wave from 4 V down to 0 V, but there is a resistor divider between the 10-Ω source resistance and the transmission line. As a result, only $50/(10 + 50) = 5/6$ of the 4 V drop actually reaches the line. 4 V*5/6 = 3.333 Volts. So at $t = 0$, a −3.333-V wave is launched down the line.

At $t = T_0$, the −3.333 V wave hits the load. A wave of 1*−3.333 = −3.333 V is reflected onto the line and a wave of 2*−3.333 = −6.666 V is transmitted out of the load.

At $t = 2\,T_0$, the −3.333 V wave hits the source. A wave of −2/3*−3.333 = +2.222 V is reflected onto the line and a wave of +1/3*−3.333 = −1.111 V is transmitted out of the source.

At $t = 3\,T_0$, the +2.222 V wave hits the load. A wave of 1* + 2.222 = +2.222 V is reflected onto the line and a wave of 2*2.222 = 4.444 V is transmitted out of the load.

At $t = 4\,T_0$, the +2.222 V wave hits the source. A wave of −2/3*2.222 = −1.481 V is reflected onto the line and a wave of +1/3*2.222 = 0.741 V is transmitted out of the source.

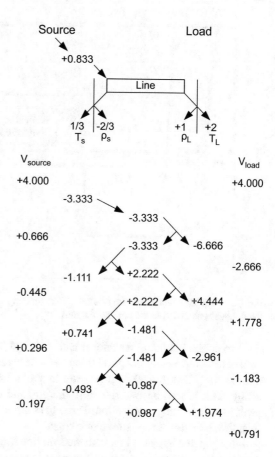

Fig. 7.2 Bounce diagram using reflection and transmission coefficients

At t = 5 T_0, the −1.481 V wave hits the load. A wave of 1*−1.481 = −1.481 V is reflected onto the line and a wave of 2*−1.481 = −2.961 V is transmitted out of the load.

And so on...

So tallying at the load, the initial voltage was +4 V. At time T_0, the voltage drops to 4−6.666 = −2.666 V. At time $3T_0$, the voltage rises to −2.666 + 4.444 = 1.778 V. At time $5T_0$, the voltage drops to 1.776−2.961 = −1.183 V, and so on.

All of this is summarized in the "bounce diagram" in Fig. 7.2.

So what happened in the example? The load voltage bounced up and down; it plunged to −2.7 V and then rose to +1.8 V and then dipped to −1.2 V, and so on. In fact, the load voltage looked *just like ringing*. However, the cause of the waveform has nothing whatsoever to do with inductance (except for the indirect connection between inductance and Z_0). The ringing-like behavior is caused by an *impedance mismatch*. A ringing-like reflection waveform is only possible if the source and load reflection coefficients have opposite signs, and is only significant if the product of the two is close to 1 (otherwise it damps out rapidly) (Fig. 7.3).

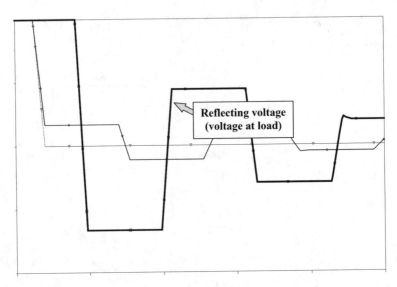

Fig. 7.3 SPICE simulation of the example. The reflecting voltage is the thick line, and it matches the voltages calculated in the bounce diagram

Why is the bouncing waveform bad? Look at the load voltage – first it plunges well below 0 V, which is a good thing because clearly, the logic gate will treat it like a logic "0." Next, however, it rises up to about 1.8 V or about 44% of a full logic swing. This is well beyond the noise margin, and the receiving gate might treat the peak as an extra logic transition. If the gate input is a clock, for example, it might treat the reflection as an extra clock edge.

Finally, what happens to a transmission line if the source stops switching as time approaches infinity? Over a sufficiently long period of time, the line becomes electrically short again, and the transmission line turns back into a wire with zero ohms series resistance. It returns to a lumped circuit. This also lines up with our intuition – that given enough time, the line acts like an ordinary wire and the entire wire drops to 0 V. This will affect the DC power consumption calculations, as we will see.

Terminating Impedance

We have seen that lines that are long (long in the electrical sense) take on two spooky properties, characteristic impedance and reflections. Is there a way to eliminate reflections altogether?

From the reflection-coefficient formula (Eq. 7.6) it is clear that if the load impedance exactly matches the line impedance, there is no reflection. This is called *load termination*. It is a parallel termination because the load is positioned as a shunt resistance from the line to ground.

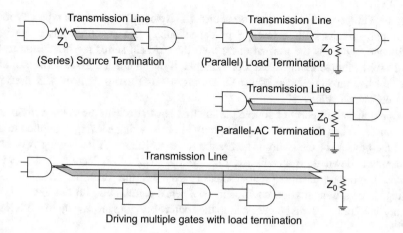

Fig. 7.4 Several termination strategies

Similarly, if the load is left open-circuited and the source impedance matches the line impedance, there is one big reflection off the load; it travels back to the source and then no more reflections. This is called *source termination*, and it (the source termination) is placed in series between the source (the driving logic gate) and the line. These are illustrated in Fig. 7.4.

These two possibilities are the two main types of termination, and they have some interesting trade-offs.

Parallel-load termination places a total resistance equal to the characteristic impedance in between the end of the line (the load end) and ground, in parallel with the logic input at the load. Typically, the source impedance is extremely low. The source reflection coefficient is close to -1, but the load reflection coefficient is close to 0 (and the load transmission coefficient is close to 1) so it does not matter. Because the resistor divider between the source resistor and the line is so lopsided, almost a full logic swing is launched into the line, and is transmitted to the load.

One disadvantage of load termination is DC power consumption. The source has to drive the load termination continuously in the steady state, since the load termination is in parallel with the logic input. One alternative to classic parallel termination, therefore, is parallel-AC termination in which a small capacitor is added in series with the resistor. This blocks the DC consumption but at AC, it still appears to be correctly terminated.

One advantage of load termination is that the load termination can be placed past the end of all of the logic inputs. It is commonly the case that one line drives multiple logic inputs, and the line can then continue past the last input and then be terminated. This is how DDR3 logic is terminated, for example, and it has the interesting (and important) advantage that the line is correctly terminated no matter how many RAM cards are inserted.

Series-source termination places a total resistance equal to the characteristic impedance in series between the source and the line, and the load is typically a very high impedance, almost an open circuit. (Again, CMOS inputs typically look like open circuits with a few pF of capacitance.) If the source has a nonzero resistance (which is the normal case) then the external series resistor needs to be reduced in value to compensate. For example, if the source has $10 \, \Omega$ resistance and the line has an impedance of $50 \, \Omega$, adding a $40\text{-}\Omega$ series resistor will create a "perfect" termination.

One unusual property of source termination has to do with the voltage divider at the source. When the source switches, only ½ of the voltage swing is applied to the line; the other ½ of the swing is across the series resistance. A half-swing is what is transmitted down the transmission line. At the load, the reflection coefficient is almost 1 and the transmission coefficient is almost 2 (like Example 7.1). The half-swing times a transmission coefficient of 2 sends a full swing to the load. Meanwhile, another half-swing races back to the source, bringing the line up to a full logic swing.

At the load, the termination is perfect – it sees one full logic swing and no reflections. At any point on the transmission line other than the load, however, there is a half swing followed, later, by another half. This is bad if there are additional logic inputs mid-way down the line, so series termination does not work well if the line drives multiple inputs, unless special provisions are made.

So what are the advantages? First, the DC power consumption is zero. Once the load has been raised to a full logic level, the voltage across the series resistor is zero and there is no current draw. Second, no special termination is needed at the load, and so the line can be extended as long as needed. PCI uses this strategy, relying on the reflection off of the open circuit to obtain a full logic swing.

When the advantages and disadvantages are traded off, parallel-load termination is the more common of the two. The half-swing issue turns out to be a deal-breaker for most logic families. And with respect to DC consumption, most modern high-speed serial interfaces are "always on" (some sort of data is always being transmitted unless the interface is completely powered down). There is continuous AC consumption, a regime in which both source and load termination have the same power consumption. Most modern high-speed signal standards operate at very low voltages, which lets logic gates switch faster and also reduces power consumption.

So how is a $50\text{-}\Omega$ line, for example, laid out correctly? From the logic output, the line is routed as close as possible to each of the inputs where the signal is being routed. By moving the line as close as possible, the "stub" that connects the line to the logic input is made as small as possible, which minimizes the load capacitance (the stub forms a capacitor with the ground plane) and keeps the stub as close as possible to a lumped element. If the stub is long enough, it becomes another transmission line, and the resulting reflections ruin the signal integrity.

After the last logic input, the line continues past the end and is then terminated with the resistor. This makes the last logic input look like a small, lumped capacitor and places a $50\text{-}\Omega$ resistance at the end. The termination is purely resistive (no capacitor in parallel) and correctly matched to Z_0, which minimizes the reflection coefficient.

It turns out that this ideal situation cannot always be achieved, as we will see in the upcoming Engineer's Notebook, but before we read about it, we need to see what happens on less-than-ideal transmission lines.

Departures from the Ideal

A transmission line is a stable, easily fabricated structure that can carry signals well up into the 10's of GHz. The key to perfect signal integrity is for the transmission line structure to have a completely uniform cross-section down its entire length.

So what causes departures from this ideal? Many things – including slots in the ground plane, vias (passages from one layer to another), right-angle bends or turns, and additional logic inputs in the middle of the line – can cause changes to the uniformity.

Most obstructions, such as vias, bends, and additional inputs, cause extra capacitance. A shunt capacitance lowers the Z_0 at the point where it connects to the line, and so causes a small negative reflection. It can be modeled as a break between two transmission lines. At that point, the first line "sees" the second line in parallel with a capacitor to ground. The capacitor is in parallel and so lowers the net impedance seen by the first line. A drop in impedance results in a negative reflection coefficient.

Since the capacitor has a derivative relationship between current and voltage, the reflection is the first derivative of the incoming waveform. For example, if the waveform is a rising edge, the reflection is a negative pulse.

These negative pulses can propagate back toward the source and look like false logic transitions at inputs that are located nearer the source. Also, if load termination is used, the pulses can reflect off of the poorly terminated source and travel back out onto the line.

In the transmission direction, the combination of the capacitance and the Z_0 of the line actually creates an RC filter that slows the signal and increases the rise time. At the point the capacitor contacts the transmission line, the Z_0 of the upstream and downstream directions are in parallel, so the effective resistance "seen" by the capacitor is $Z_0/2$. The overall rise time adder is $2.2 * Z_0/2 * C = 1.1 Z_0 C$. If the signal passes multiple capacitive loads (a common case on a bus), then the rise time steadily grows as the signal moves down the line (and a stream of negative pulses is sent back to the source).

A SPICE simulation of a capacitive imperfection is shown below in Fig. 7.5. The circuit in the simulation used source termination, so the initial reflection of the falling edge off of the capacitor is absorbed at the source. However, in source termination, there is a nearly total reflection off of the load. The load reflection actually reflects off of the capacitor and comes back to the load. Since the sign of the capacitor's reflection coefficient is negative, the reflection of the falling edge is a rising spike. Since the sign is opposite, it can be mistaken by the load as an extra logic transition. The simulation also shows how the "reflection of the reflection" turns into a second-derivative double pulse. (The voltage at the load is drawn in the thick line in Fig. 7.5.)

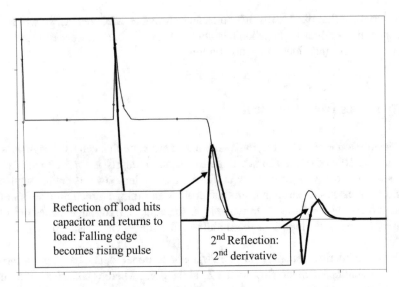

Reflection off load hits capacitor and returns to load: Falling edge becomes rising pulse

2nd Reflection: 2nd derivative

Fig. 7.5 Capacitor in middle of the line with source termination

To recap, the source sent a half-logic-swing waveform down the line and past the capacitor. The half-logic-swing reflects off of the load and heads back to the source. When the reflection hits the capacitor, there is a reflection back to the load, which is the first pulse seen in the simulation. Additional reflections also occur later.

If there is a slot or hole in the ground plane, the return current has to divert around it, and the effect is an increase in inductance at that point in the line. It looks, electrically, like a small series inductor is interrupting the transmission line. It raises Z_0 and so causes a positive first-derivative reflection. For example, a rising edge is reflected as a positive pulse. The positive pulses are not a problem (they overdrive the incoming signal and do not look like a false logic transition) but the pulses reflected off of the source are negative and can cause problems. Like their capacitive brethren, the inductive imperfections are low-pass filters in the forward direction with a rise time adder of 1.1 L/Z_0.

A SPICE simulation of an inductive imperfection is shown below in Fig. 7.6. This circuit used load termination, so the positive reflection off of the inductor travels back to the source, hits the source's negative reflection coefficient, and heads back down to the load. So the falling edge created a falling spike which, after reflecting off of the source, became a rising spike.

To recap, a falling edge was launched down the line. The edge reflected off of an inductor (and became a pulse in the process) and went back to the source. The source then reflected the pulse back out onto the line and back toward the load. This is the first pulse visible in the simulation.

These capacitive and inductive issues highlight the general process of obtaining good signal integrity – minimize vias, right-angle turns, and slots or cuts in the ground plane. One common way to reduce the effect of right-angle turns is to turn corners gradually; come circuit-board layout software will let you lay it out this way.

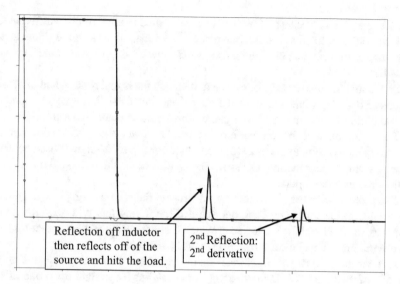

Fig. 7.6 SPICE simulation of an inductive imperfection

Fixing It When It's Broke

In most designs, termination tends to be straightforward. More problematic is trying to keep lines straight, on the same circuit-board layer, and avoiding interruptions in the power or ground plane. The problem arises because circuit boards are so crowded that often signals have to take detours. The straight line from chip A to chip B might be right under chip C, for example, or chip B might be on the opposite side of the board.

The "trick" is to keep track of the high-speed buses and pay special attention to their layout. We did not, and we got burned...

Engineer's Notebook: The High-Speed Bus Disaster
Two of the main chips in one of our designs were connected by a very high-speed parallel bus (16 bits wide, roughly 500 MHz speed). Parallel buses are automatically harder to design because, in order for all the bits to stay in sync, each bit must have the same length.

On our first version of the board, the bus was a mess, full of errors. Then we sat down and looked at the layout, and it was embarrassing. We had made one major rookie mistake and discovered a subtle issue that kept us from being able to follow the rules.

The bus had to be routed on the bottom side of the board. This turns out to be an important distinction because, as is typical in a four-layer-board design, the bottom side is routed over a power plane instead of over a ground plane. (You can add almost as many planes as you want; it just gets more expensive, and our designs were very cost-sensitive.) The power plane was a hodge-podge of voltages, including 1.0,

1.2, 1.8, and 3.3 V, to name a few. The bus was routed over all of these different DC voltages, and had to cross a slot in the plane every time. It was as if the signal had to drive across 10 different countries and stop and show a passport every time it crossed a border.

So the first fix was to carve out space underneath the signal routing and fill it with a dedicated 0 V ground plane all the way from under the first chip to under the second chip. The bus then had a single, uniform ground plane under it.

Then we looked at the termination. We had "followed the rules" and routed the signals under the chip, to the input pins, and then continued the signals back out from under the chip to terminating resistors. That is, we located the termination at the end of the line, past the inputs.

The problem was the signal routing underneath the chip. Having a wire go from the outside to the input and back out doubled the wiring that was needed, and the chip was so small and the wiring so crowded that the wiring was a tangled-up mess (to use the precise technical term). So we broke the rules.

As near the chip as we could place them, we added terminating resistors on the top side of the board. We then ran a via from each bus bit straight up to one side of the resistor and then ran from the other side of the resistor through a via to the top-side ground plane. In other words, we had nearly perfect termination wiring. Then we continued the bus on the bottom side under the chip. The wire lengths past the terminating resistors were pretty long, over an inch in some cases, but it actually resulted in much cleaner termination.

After all those changes, we had error-free data transmission.

Appendix

1. Reconstruct the transmission-line Example 7.1 from the chapter. The example was 8″ of microstrip with $Z_0 = 50\ \Omega$ and $\varepsilon_r = 5$.

   (a) Compute the T_P (ns/ft), C_L (pF/ft), and T_0 (one-way propagation delay) of the line.
   (b) Run a Spice simulation of the line. The source has $10\ \Omega$ of impedance and the load has $5\ M\Omega$ of impedance. The source is a 4–0 V transition with a 500 ps fall time.
   (c) What is the undershoot and overshoot seen at the load.

      (i) At the first arrival of the falling edge
      (ii) At the arrival of the first reflection

   (d) Change the load termination to $50\ \Omega$, and rerun the simulation.

      (i) What is the waveform seen at the load?
      (ii) What is the waveform seen at the source?

2. Reconstruct the transmission-line example (6.4) from the lecture (p. 29 from the PowerPoint). The example was 8″ of microstrip with $Z_0 = 50\ \Omega$ and $\varepsilon_r = 5$.

   (a) Compute the Tp (ns/ft), C_L (pF/ft), and T_0 (one-way propagation delay) of the line. The Tp formula is in the PowerPoint slides, and recall that $C_L = T_p/Z_0$ and that $T_0 = T_p \times$ length of line. $Tp = 85*\text{sqrt}(0.475*5 + 0.67) = 148.3$ ps/in $= 1.779$ ns/ft. CL $= Tp/Z0 = 1779$ ps/ft./50 $= 35.6$ pF/ft. $T_0 = 1.779$ ns/ft. $* 8/12$ feet $= 1.186$ ns.

   (b) Run a Spice simulation of the line. The source has $10\ \Omega$ of impedance and the load has $5\ \text{M}\Omega$ of impedance. The source is a 4–0 V transition with a 500 ps fall time.

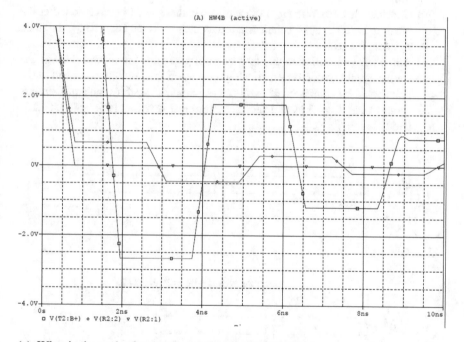

(A) HW4B (active)

   (c) What is the undershoot and overshoot seen at the load.

      (i) At the first arrival of the falling edge -2.666 V
      (ii) At the arrival of the first reflection $+1.777$ V

   (d) Change the load termination from 5 MΩ to 50 Ω, and rerun the simulation.

      (i) What is the waveform seen at the load? Falling edge
      (ii) What is the waveform seen at the source? Falling edge

3. Construct a SPICE simulation of a transmission line with $Z_0 = 50 \ \Omega$ and $T_0 = 2$ ns. The source is a 1–0 V transition with a 200 ps fall time. The source has 5 Ω impedance, and the load impedance is 50 Ω in parallel with 5 pF capacitance. Then split the line in half (i.e., create 2 50-Ω transmission lines in series each with $T_0 = 1$ ns) and simulate the following issues in the middle of the line. Plot the voltage at the source, in the middle of the line, and at the load for each of the 4 cases.

(a) Add a logic gate input in the middle of the line (6 pF total capacitance – 5 pF gate plus 1 pF via). The extra logic gate does not have a 50-Ω terminating impedance. (Why?) Simulate in SPICE.

(b) Add a break in the signal-return plane (5 nH inductance) and simulate in SPICE.

(c) Consider case b. Compute a value of capacitance to add before and after the break to compensate for the extra inductance. Add the capacitance and then simulate it again.

(d) Add a third transmission line by placing it parallel to the second. That is, the source should be connected to one transmission line, and the other end of the line should be connected to two lines in parallel. The third transmission line should have $T_0 = 2$ ns and should be terminated with a logic gate (50 Ω in parallel with 5 pF).

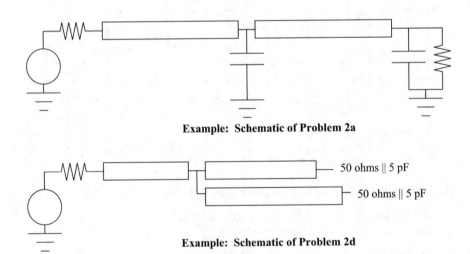

Example: Schematic of Problem 2a

Example: Schematic of Problem 2d

4. You are helping a coworker debug a nonfunctioning board. One signal trace has a one-way propagation delay of $T_0 = 4$ ns. You and the coworker probe the signal source. 2 ns after the source creates a *rising* edge, there is a large *negative* spike visible on the oscilloscope.

   (a) What is the likeliest cause for the spike?
   (b) The signal trace is about 27 inches long. How far down the trace would you expect to find the cause of the spike?

Chapter 8
Lossy Transmission Lines

Background and Objectives

In Chaps. 6 and 7, the concepts of a transmission line, reflections, and terminations were unveiled. Transmission lines have the very desirable property that they carry signals nearly completely intact, and that is why they are widely used. However, if you go all the way back to the beginning of Chap. 6, one assumption was made: That is, in a coaxial cable or on a twisted-pair cable, there is no loss of signal energy. This turns out to be false, but the source of the loss is surprising. This chapter discusses dielectric loss, explains how it manifests itself, and explains how it affects signals on real circuit boards. When this chapter is finished, you should be able to:

- Understand the units of attenuation
- Contrast frequency-independent and frequency-dependent attenuation
- Identify the two sources of frequency-dependent attenuation
- Explain dielectric loss and its physical origins
- Explain how dielectric loss affects ε
- Show how to read dielectric loss off of a datasheet
- Describe how dielectric loss affects signal propagation

What Is Attenuation and How Is It Measured?

This chapter is ultimately about attenuation, which is a loss of intensity. In the context of signal integrity, attenuation is when the energy of a signal becomes weaker, usually over some distance or space.

One common source of attenuation is the propagation of a signal through a three-dimensional medium. Radio waves, for example, become weaker at the square of the distance. The waves are "spreading out" to fill space, spreading out over larger areas, and therefore the energy per unit area drops.

© Springer Nature Switzerland AG 2022
S. H. Russ, *Signal Integrity*, https://doi.org/10.1007/978-3-030-86927-4_8

If the signal energy is confined, as is the case with a transmission line, the signal energy should remain constant but, in fact, it also drops over time. Every "real" transmission line has attenuation.

But first... How is attenuation measured? Almost all attenuations are exponential. For example, over a distance of 1 m, a signal may be reduced in amplitude by ½. So if a signal has an amplitude of 8 V at the source it may be reduced to 4 V at 1 m, 2 V at 2 m, and so on. Over a distance of 1 m, the signal is reduced by a factor of 2, which corresponds to a 6 dB drop in amplitude. So the attenuation is 6 dB/ 1 m = 6 dB/m. This unit of measure (dB per distance) is a common measure of attenuation, and reflects the exponential nature of attenuation.

Example 8.1

A cable has an attenuation of 0.1 dB/m.

(a) What is the amplitude of a 5 V signal after traveling down 6 m of this cable?

Attenuation = 0.1 dB/m * 6 m = 0.6 dB.
0.6 dB of loss = $10^{-(0.6/20)}$ = 0.933.
Output = 5 * 0.933 = 4.665 V.

(b) At what distance does the signal have an amplitude of 4 V?

4 V/5 V = 0.8.
$10^{(-x/20)}$ = 0.8; $-x/20$ = −0.9691; x = 1.9382 dB of loss.
Distance = 1.9382 dB/0.1 dB/m = 19.4 m.

The next issue to consider is whether the attenuation is constant with frequency or varies with frequency. This turns out to be a profound issue.

If the attenuation is constant with frequency, then the attenuation causes the signal to shrink linearly. Most importantly, the rise and fall times are not degraded and so the signal retains very sharp edges. While the receiver may need to use amplification, the basic process of receiving and decoding the signal is relatively simple because the shape of the signal has not been altered. The receiver will eventually reach a point where the amplifier is not enough, but this still leads to a robust system. For example, a Wi-Fi receiver may have a dynamic range (also known as link margin) of 100 dB – five orders of magnitude. (Actually, Wi-Fi is a bad example – it operates at a specific frequency and so all of the signal is attenuated the same amount, even if there is frequency-dependent attenuation present.)

If the attenuation varies with frequency, specifically if the attenuation is worse at high frequency, this leads to significant signal distortion. Portions of a signal with high-frequency content, such as rising and falling edges, are attenuated more than the lower-frequency counterparts. This manifests itself as a signal with significant rise time degradation, which is the same as saying a signal that has been low-pass filtered. On an oscilloscope, for example, the relatively square-ish rising and falling signals become much rounder, and the receiver has a much harder time sampling whether the input is 0 or 1.

The contrast is illustrated below in Fig. 8.1.

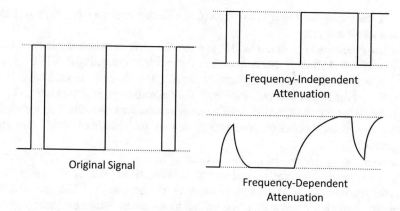

Frequency-Independent
Attenuation

Original Signal

Frequency-Dependent
Attenuation

Fig. 8.1 Frequency-independent and frequency-dependent attenuation

Frequency-Dependent Attenuation

The first example of frequency-dependent attenuation is one we have already studied, the skin effect. Because of this effect (which is due to inductance, see Chap. 4), the conductor appears to get smaller and thinner as frequency goes up. It is as if the resistance of the conductor is going up. As you may recall from the discussion, the skin depth goes down (gets smaller) as the square root of frequency, and so the effect of skin depth is that the apparent resistance of a conductor goes up as the square root of frequency.

The frequency region below which the skin depth is greater than half the conductor thickness is called the "low loss" region because, in those frequency ranges, the signal can spread out over the entire conductor and the resistance is constant with frequency. The frequency region above is called the "skin effect" region, and the resistance goes up with the square root of frequency.

However, once the frequency becomes even higher, another effect appears.

Dielectric Loss

Think about this: If you put an empty glass bowl in a microwave oven and run the microwave for a few minutes, will it get hot? The answer is, of course, is "yes." But why? Microwave ovens work by emitting microwaves at a frequency that water can absorb (2.4 GHz). But there is no water in the glass. So how does it get hot?

Backing up for a moment, what is the significance of the bowl getting hot? It means that something in the bowl is absorbing real energy (not imaginary energy) from the microwaves in the oven. So what is it?

It turns out that molecules have microscopic regions of positive and negative charges, due basically to the fact that they are asymmetric. When an electric field is applied, the molecules rotate to line up with the field. Even in a solid, the molecules rotate a small amount.

When an alternating electric field is applied, the molecules actually rock or twist back and forth. At microwave frequencies, the incident alternating E field begins to match the speed at which the molecules move. Even though the molecules twist really fast, if the E field is alternating faster, the molecules never stop moving. As the frequency goes up, the speed of rotation of the molecules goes up. The molecules undergo an actual, physical, mechanical motion that becomes faster at higher frequencies.

Conversely, the molecules require real energy to do this. Actual work (force times distance) is being done, and so real energy is being consumed. So what is another term for the microscopic vibration of molecules? Temperature! The result of this microscopic motion is that the molecules, and therefore the entire substance (e.g., the entire glass bowl), get hot.

Since transmitting microwaves through insulating materials is accompanied by the loss of real energy, the name for this phenomenon is *dielectric loss*.

How Does Dielectric Loss Affect ε?

So if one applies high frequencies (e.g., GHz) to a capacitor, two different phenomena are observed.

The first is that, for a fixed voltage, the current goes up with frequency, and this current is 90° out of phase with the voltage. This is the current we are all familiar with, a manifestation of $I = C dV/dt$.

The second is that, for a fixed voltage, the current also goes up with frequency, but this current is *in phase* with the voltage. This is the current due to dielectric loss. It is in phase with the voltage because it is real energy, and it is real energy because real work is being done and being dissipated as heat. It is proportional to frequency because higher frequencies twist the molecules twice as fast.

Pause for a second: Note that both currents are linearly proportional to the voltage V and to the frequency ω.

For the analysis, it is convenient to flip everything around and speak in terms of admittance. So instead of $V = IR$ we will use $I = VG$. The reason for this will become clear soon.

For an AC voltage V at frequency ω, we have

$$I = j\omega CV + \omega \widehat{C} V \tag{8.1}$$

The first term is the familiar one, the 90-degree-out-of-phase current. The second is due to this new phenomenon, dielectric loss. In Eq. 8.1, the real part of the current is being treated as if it is a new type of capacitance. This abstraction works because the real current behaves like a capacitance (conductance increasing with frequency,

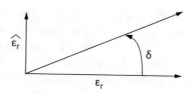

Fig. 8.2 Real and imaginary part of dielectric constant

current proportional to frequency). In fact, if one measures dielectric loss in a lab, it becomes apparent it is an intrinsic property of the material.

One can convert the "C" term above into two terms, and so

$$C \equiv \varepsilon_r C_0 \tag{8.2}$$

where C_0 is the *intrinsic capacitance*. The intrinsic capacitance is the capacitance that would be there if the dielectric were replaced with a vacuum. A very similar substitution can be made for the dielectric-loss term $\widehat{C}$, and so

$$I = j\omega\varepsilon_r C_0 V + \omega\widehat{\varepsilon}_r C_0 V = \omega C_0 V(\, j\varepsilon_r + \widehat{\varepsilon}_r) \tag{8.3}$$

If you look at the term in parentheses, it becomes clear that *the relative dielectric constant is now a complex number*! This is where it gets very confusing, so pay attention.

The convention is to express $\widehat{\varepsilon}_r$ in terms of its ratio to ε_r. In order to compute this ratio, one treats the combination of $\widehat{\varepsilon}_r$ and ε_r as a complex number, with ε_r being the real part and $\widehat{\varepsilon}_r$ as the imaginary part. So ε_r, which is associated with the *imaginary* component of the current, is treated as the *real part* of the dielectric constant. Conversely, $\widehat{\varepsilon}_r$, which is associated with the real component of the current, is treated as the imaginary part of the dielectric constant. It sounds backwards, but that is the convention. It dates back to the fact that the imaginary current/real dielectric constant association came along first, and the discovery of dielectric loss later.

One can observe the dielectric constant in the Argand plane (Fig. 8.2).

The phase angle of the now-complex dielectric constant is termed δ and so we have

$$\tan \delta = \widehat{\varepsilon}_r / \varepsilon_r \tag{8.4}$$

From (8.4), it can be seen that $\widehat{\varepsilon}_r$ can be expressed as $\widehat{\varepsilon}_r = \varepsilon_r \tan \delta$ and so

$$I = \omega C_0 V(\, j\varepsilon_r + \widehat{\varepsilon}_r) = \omega C_0 V(\, j\varepsilon_r + \varepsilon_r \tan \delta) = \omega\varepsilon_r C_0 V(\, j + \tan \delta) \tag{8.5}$$

Since $C = \varepsilon_r C_0$,

$$I = \omega C V(\, j + \tan \delta) \tag{8.6}$$

The real part of I/V, therefore, is $\omega C \tan \delta$. Also note that the real part of I/V is conductance. (And, to recap, the imaginary part is the current one normally associates with a capacitor, 90 degrees out of phase with the voltage.)

To review, real dielectrics exhibit a phenomenon called dielectric loss, which causes the dielectric to absorb real energy. It is intrinsic to the material and can be expressed as a ratio relative to the ε_r of the material. The ratio is called $\tan \delta$, and can be thought of as the imaginary component of the dielectric constant of the material.

So how big is $\tan \delta$? For example, the $\tan \delta$ for FR-4 fiberglass (the most common circuit-board material) is roughly 0.02.

How do you know $\tan \delta$? It is normally listed on a datasheet or in reputable online sources. So dielectric loss is measured in the relatively obscure unit of $\tan \delta$, and this unit is readily available.

How Does Dielectric Loss Affect Signal Propagation?

The basic assumption of a transmission line was that the conductance of the insulator was zero and the resistance of the conductors was zero. If the conductance is nonzero, then the model breaks down, and we have to go back to the lumped model of a transmission line (see Fig. 6.3) to figure out what actually happens.

Before, in the approximate model of a transmission line, the assumption was that $R = 0$ and $G = 0$. The assumption was that the conductors were perfectly conductive and the insulator was perfectly insulating. Here, because of dielectric loss, we discover that $G > 0$. However, we will still assume that the resistance R is zero; this assumption turns out to be accurate because the resistance is much smaller than Z_0.

If one considers the circuit from the point of view of the conductance G (again, see Fig. 6.3), it is apparent that there is a resistor divider. One piece of the approximate model is reproduced in Fig. 8.3, to help understand the math.

The conductance G "sees" the series resistance of Z_0 on its input and output, and so the net series resistance is $Z_0/2$. The inductance, capacitance, and resistance R are shown as grayed out in the picture – we are neglecting them because we are considering the pure DC behavior of the circuit and, as noted above, R is approximately 0. If a voltage V_{in} is applied, then the voltage V_{out} is the resistor divider

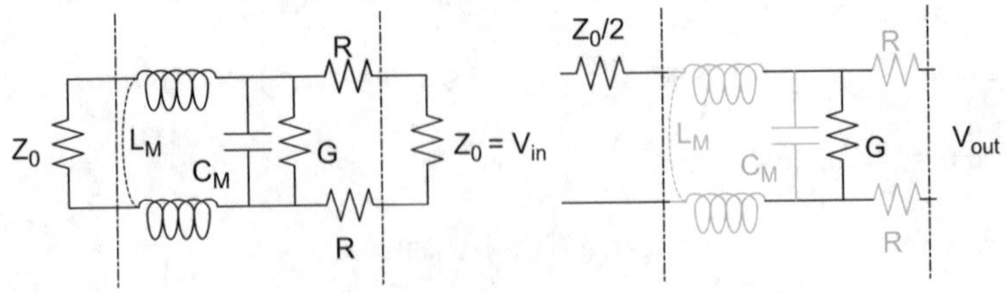

Fig. 8.3 One "piece" of a transmission line, with the other pieces replaced with Z_0

formed by the conductance G (actually, by the resistance $1/G$) and $Z_0/2$. Also, we can express G as a conductance per unit length G_L, which is the convention when we divide the transmission line into little pieces. So we have

$$V_{out} = V_{in} \left(\frac{\frac{1}{G_L}}{\frac{1}{G_L} + \frac{Z_0}{2}} \right) = V_{in} \left(\frac{1}{1 + \frac{1}{2} G_L Z_0} \right) \tag{8.7}$$

We will need to convert this ratio to an exponential form. To make it solvable, we start by expressing the ratio as an exponent of e.

$$\frac{V_{out}}{V_{in}} = \frac{1}{1 + \frac{1}{2} G_L Z_0} = e^{-x} \tag{8.8}$$

$$\ln \left(\frac{1}{1 + \frac{1}{2} G_L Z_0} \right) = -x \tag{8.9}$$

$$\ln \left(1 + \frac{1}{2} G_L Z_0 \right) = x \tag{8.10}$$

For the next step, we can use the approximation $\ln(1 + a) \approx a$ because, in most practical cases, $G_L Z_0$ is a very small number and so $G_L Z_0{}^2$ is much less than $G_L Z_0$. So

$$x \approx \frac{1}{2} G_L Z_0 \tag{8.11}$$

and

$$\frac{V_{out}}{V_{in}} = e^{-x} = e^{-\frac{1}{2} G_L Z_0} \tag{8.12}$$

Converting this to dB is straightforward. For a loss b in dB, we can write

$$e^{-\frac{1}{2} G_L Z_0} = 10^{-(b/20)} \tag{8.13}$$

$$\log_{10} \left(e^{-\frac{1}{2} G_L Z_0} \right) = \log_{10} \left(10^{-(b/20)} \right) \tag{8.14}$$

$$-\frac{1}{2} G_L Z_0 \log_{10}(e) = -b/20 \tag{8.15}$$

$$10 G_L Z_0 \log_{10}(e) = b \tag{8.16}$$

Since $\log_{10}(e) = 0.434$,

$$b = 4.34 G_L Z_0 \tag{8.17}$$

where b is in units of dB/length. We can express both G_L and Z_0 in terms of their per-length parameters. Equation 8.6 above teaches that $I = \omega C V \tan \delta$. Since $G_L = I/V$ and C is C_L (capacitance per length), we have that $G_L = \omega C_L \tan \delta$. From Chap. 6, we have $Z_0 = \sqrt{\varepsilon_r}/cC_L$. (And, yes, c is the speed of light.) So b, the loss in dB, can be written as

$$b = 4.34 G_L Z_0 = 4.34 \frac{\omega C_L \tan \delta \sqrt{\varepsilon_r}}{c C_L} = \frac{4.34 \omega \tan \delta \sqrt{\varepsilon_r}}{c} \tag{8.18}$$

where b is the loss in dB/length. The units of length are whatever unit of length is used for c; all other measurements cancel out or are dimensionless.

Example 8.2

Consider FR-4 fiberglass with $\varepsilon_r = 4.2$ and $\tan\delta = 0.02$.

(a) Find the loss in dB/cm due to dielectric loss at a frequency of 10 GHz.

Recall that c, the speed of light, is 300,000 km/s $= 3 \times 10^{10}$ cm/s. (We need c to be in cm/s so that the final answer is in dB/cm.) So we have $b = (4.34 * 2 * \pi *$ 10 GHz $* 0.02 * \sqrt{4.2})/3 \times 10^{10}$ cm/s $= 0.372$ dB/cm.

(b) Over what length of conductor will a 10 GHz signal have 3 dB of loss?

3 dB/0.372 dB/cm $= 8.05$ cm.

The example shows that dielectric loss can significantly reduce GHz-range signals in FR-4 fiberglass.

Is there an easier way to understand the effect? First, recall that t_r is roughly equal to $0.35/f_{3dB}$. That is, the rise time of a circuit element is 0.35 divided by its 3 dB frequency (the frequency at which the element attenuates its input by 3 dB). Second, we can re-write (8.17) by multiplying the loss in dB/length by length (resulting in a loss in dB) and setting that loss equal to 3 (that is, 3 dB). So we have, for distance d,

$$3 = \frac{4.34 \omega \tan \delta \sqrt{\varepsilon_r}}{c} d = \frac{4.34 \cdot 2\pi f_{3\ dB} \tan \delta \sqrt{\varepsilon_r}}{c} d$$

$$= \frac{27.26 f_{3\ dB} \tan \delta \sqrt{\varepsilon_r}}{c} d \tag{8.19}$$

Solving for $f_{3\ dB}$

$$f_{3\ dB} = \frac{3c}{27.26 \tan \delta \sqrt{\varepsilon_r} d} \tag{8.20}$$

Substituting (8.19) into the rise time formula, we have

$$t_r = \frac{0.35}{f_{3\,dB}} = 0.35\frac{27.26 \tan \delta \sqrt{\varepsilon_r} d}{3c} = \frac{3.18 \tan \delta \sqrt{\varepsilon_r} d}{c} \qquad (8.21)$$

Example 8.3
Continue Example 8.2 with FR-4 fiberglass with $\varepsilon_r = 4.2$ and $\tan\delta = 0.02$.

(a) Find the rise time adder of 1 cm of conductor routed over FR-4 due to dielectric loss.

Again, c is 3×10^{10} cm/s. So we have $t_r = 3.18 * 0.02 * \sqrt{4.2} * 1$ cm/3×10^{10} cm/s $= 0.434$ ps. This is a very important figure of merit: FR-4 has a rise time adder of 0.434 ps/cm due to dielectric loss. Stated differently, every cm of FR-4 is a low-pass filter with a rise time equivalent to a 0.434 ps rise time adder.

(b) A 1 Gigabit/s SATA signal with a rise time of 130 ps is routed over 10 cm of FR-4. What is its rise time at the load?

Total rise time adder $= 0.434$ ps/cm $* 10$ cm $= 4.34$ ps. $\sqrt{130^2 + 4.34^2} = 130.1$ ps. This is not a significant issue.

(c) A 6 Gigabit/s SATA signal with a rise time of 20 ps is routed over 30 cm of FR-4. What is its rise time at the load?

Total rise time $= 0.434$ ps/cm $* 30$ cm $= 13.02$ ps. $\sqrt{20^2 + 13.02^2} = 23.9$ ps. Since this is a 20% increase in rise time, this is a significant problem.

(d) How much shorter does the trace need to be to keep the rise time under 22 ps?

$\sqrt{20^2 + x^2} = 22$ ps. Solving, we have $x = 9.17$ ps. 9.17 ps/0.434 ps/cm $= 21.1$ cm.

This example teaches two things.
First, the dielectric loss ultimately manifests itself as a rise time adder. The units can be a little confusing, but the dielectric loss in FR-4 creates a rise time of 0.434 ps per cm of trace length. The trace length in cm times 4.34 is the rise time adder in ps. This rise time must then be combined with the rise time of the signal (using the rise time adder formula) in order to calculate the rise time of the load.
Second, this example shows that, as signals move past 1 Gbit/s, FR-4 fiberglass is going to be harder to use. Because of the combination of dielectric loss and high-speed signaling, the era of FR-4 is nearing its end.

Appendix

1. A circuit board trace has an attenuation of 0.3 dB/m and is carrying a waveform that is 1.2 V peak-to-peak.

   (a) What is the amplitude of the signal after traveling down 30 cm of trace?
   (b) What attenuation would be needed to lose less than 10% of the peak-to-peak voltage at a distance of 30 cm?

2. Consider a PTFE ceramic circuit-board material (Rogers R3003™) with $\varepsilon_r = 3.0$ and $\tan\delta = 0.0013$.

   (a) Find the loss in dB/cm due to dielectric loss at a frequency of 7 GHz.
   (b) Over what length will a conductor with a 10 GHz signal have 3 dB loss?
   (c) How does this compare to the FR-4 fiberglass in Example 9.2?

3. Consider a material with $\varepsilon_r = 1.7$ and $\tan\delta = 0.01$.

   (a) Find the rise time adder of 1 cm of conductor routed over the material due to dielectric loss.
   (b) A microwave radio signal with a rise time of 15 ps is routed over 30 cm of the material. What is the rise time at the load?
   (c) What length does the trace need to be for a rise time of 17 ps?

4. Consider a circuit trace carrying a 3 Gbps SATA signal. It has a 250-mV logic swing. The trace is routed over a circuit board material that has a dielectric loss adder of 4 ps/cm.

   (a) If the trace is routed 10 cm on the board, what is the effective rise time of the dielectric loss? 4 ps/cm * 10 cm = 40 ps.
   (b) If the SATA signal has a rise time at the driver of 50 ps, what is the rise time of the signal at the other end of the trace? Sqrt($50^2 + 40^2$) = 64 ps.
   (c) How long can the signal be routed and keep the overall rise time (combination of driver rise time and rise time due to dielectric loss) below 60 ps? sqrt $(50^2 + x^2)$ = 60 ps. X = 33.2 ps. 33.2 ps/4 ps/cm = 8.3 cm.

Chapter 9
Differential Signaling

Background and Objectives

In exploring transmission lines and signals, it is clear that signals that are close to each other have both capacitive and inductive coupling between them. It turns out that this is both good news and bad news. In this chapter, we will start with the good news and introduce differential signaling. In the next chapter, we will break the bad news and discuss crosstalk.

One common method for better signal integrity is to move to differential signaling. This seemingly simple change, using both a positive and negative waveform in a matched set, eliminates the need for an intact ground plane, takes care of return current, manages ground bounce, and generally improves signal integrity. When this chapter is finished, you should be able to:

- Explain how differential signaling improves signal integrity.
- Calculate Z_{diff}, the differential impedance.
- Explain how changes in mutual inductance and capacitance can affect Z_{diff}.
- Terminate cables or conductors that carry differential signals.
- Simulate differential signaling.
- Explain the effects of common-mode signals and clock jitter on differential signals.

What Is Differential Signaling and How Does It Help?

Up until now, a ground plane has been used as a return path. This has one advantage; you can lay out signals over a ground plane and never have to worry about laying out a return path. But there are disadvantages. For example, the ground plane has to be kept intact, or else the inductance (both self and mutual) will go up. The ground plane is used for pretty much everything on the board (e.g., for power distribution,

© Springer Nature Switzerland AG 2022
S. H. Russ, *Signal Integrity*, https://doi.org/10.1007/978-3-030-86927-4_9

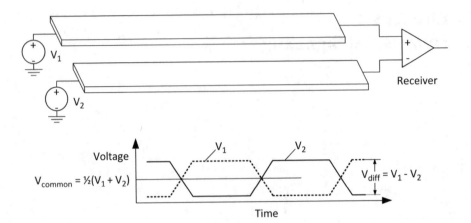

Fig. 9.1 Example of differential signaling. V_1 and V_2 are the differential signals, and their voltages over time are shown on the bottom

for return currents for unrelated signals, etc.). Is there an alternative to using a ground plane?

It turns out there is. In this alternative strategy, each signal is routed with its dual, so that when one switches high, the other signal switches low. This tends to cancel out ground bounce, and the receiver can operate off of the difference between the two signals. Most importantly, each signal now has a dedicated return-current path and so the signals stay "cleaner." It sounds strange, but by routing a signal's dual near the signal, the dual of the signal functions as a return path. Since the ground plane is no longer needed for signal propagation, the pair of signals is relatively robust if the surroundings change, such as if there is a slot in the ground plane. In fact, the ground plane can be done away with altogether.

This strategy of routing each signal with its own dual is called *differential signaling*. In fact, low-voltage differential signaling (LVDS) is taking over most modern interconnection standards. USB, PCI Express, SATA, and even the old Firewire standard all use differential signaling. It is the "trick" that enables the modern standards (like SATA) to run at Gbps over cheap cabling.

So what does it look like electrically? Consider a differential pair (sometimes informally called a "diff pair") using signals V_1 and V_2. The routing and voltage levels are illustrated below in Fig. 9.1.

The two signals are routed over a ground plane (or perhaps not – it turns out not to matter much) and go to a receiver. The receiver subtracts the two signals in order to extract the differential signal. The voltage graph in Fig. 9.1 bears some explanation. As V_1 rises, V_2 falls. Thus V_2 is always the opposite (or dual) of V_1.

V_1 is the voltage on line 1, and V_2 is the voltage on line 2. So we have

$$V_{\text{diff}} = V_1 - V_2 \tag{9.1}$$

$$V_{\text{common}} = (V_1 + V_2)/2 \tag{9.2}$$

where V_{diff} is the differential voltage and V_{common} is the common voltage. If it is not already clear, V_{diff} is the voltage that carries information, and it is defined as the difference between the two voltages. A logic 1 may be defined as the case where $V_1 > V_2$, for example. The receiver can subtract the two signals (like at the input of an op-amp), and can thereby make a precise voltage measurement. Sometimes the common voltage is zero and sometimes it is some positive DC offset. The receiver uses the differential voltage almost exclusively, and for most practical purposes the common voltage can be ignored.

There are only two situations where the common voltage becomes relevant. First, if the common voltage becomes high enough (or low enough), it saturates the receiver. (Amplifiers do not behave normally if inputs are close to the power or ground voltages.) Second, common voltage can cause unwanted radiated emissions, as will be discussed in more detail below.

So the common voltage can usually be ignored, and this is why differential signaling is so robust. Since the receiver is subtracting the two signals, the system is tolerant to noise and other signal impairments. As long as the two signals are corrupted in the same way (e.g., receive the same noise pulse), the difference between them is still intelligible at the receiver.

To summarize, the recipe for a differential pair is quite simple. First, you need two lines that are identical in length, Z_0, and layout impairments. Second, it is preferable that the lines be close enough to obtain capacitive and inductive coupling.

What Is Z_{diff}?

The characteristic impedance of the differential pair is called Z_{diff} and it turns out to be easy to estimate.

Consider the case when V_1 is rising and V_2 is falling. V_1 is sourcing a current I into its line and V_2 is sinking and equal and opposite current I from its line. This is illustrated in Fig. 9.2.

The voltage swing is doubled, because the falling edge of V_2 is subtracted from the rising edge of V_1, and the two swings are of equal magnitude. That is, the total voltage swing of V_2-V_1 is double the swing of a single line. However, the current looks (to the differential source) like a single loop of current. So the voltage is doubled but the apparent current is not.

$$Z_{\text{diff}} = {2V}/{I} = 2Z_0 \qquad (9.3)$$

Fig. 9.2 Driving a signal into the differential pair

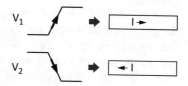

Fig. 9.3 Termination for
differential signals. Note
that each line has a
characteristic impedance of
Z_0

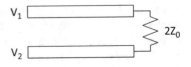

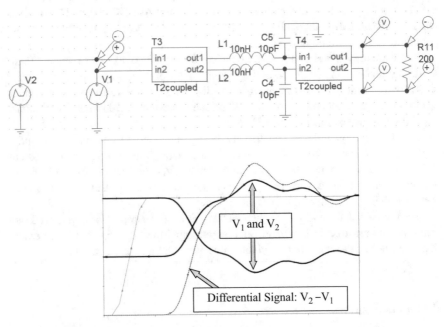

Fig. 9.4 Simulation of differential signal with inductive and capacitive impairments

So $Z_{\text{diff}} \approx 2Z_0$. This leads to the termination strategy for differential signaling
(Fig. 9.3).

So how well does it work? One can simulate a case where there is a big
inductance and big capacitance on both lines. This is shown in Fig. 9.4.

The results are well-behaved, even when one stops to consider that there is a lot of
inductance and capacitance in the circuit. The differential signal (shown in the dotted
line in Fig. 9.4) shows about a 50% overshoot, but then settles down to a normal
level fairly soon. The signals still work correctly, since the receiver is subtracting the
two signals, even though there is capacitance and inductance. This is a dramatic
improvement over conventional signaling. This example works well because inductance and capacitance are symmetric – each line has the same inductance and
capacitance.

What happens if the differential-voltage swing lines are brought closer? Consider
the situation from the point of view of V_1. Its own signal is rising, for example, and
the opposite signal is falling. V_1 has to charge its own line and, at the same time,
combat the negative current introduced from V_2 due to capacitive coupling. In other
words, it has to drive harder and source more current in order to obtain the same

voltage swing. The effect is to lower the apparent Z_{diff} "seen" by V_1 and V_2. At extremely close spacing, Z_{diff} can drop as much as 12%.

There are formulas that enable one to approximate Z_{diff} for practical cases such as microstrip and stripline. However, these formulas are very approximate, and so a field solver is often used if precision is needed.

What happens if the ground plane is moved farther from the two signal lines? As the ground plane moves farther, the coupling from the signals to the ground drops, and at some point, becomes negligible. At that point, the differential pair is carrying all of its own return current. This point is typically when the conductor-to-ground spacing is roughly double the conductor-to-conductor spacing. At that point, if it is not already clear, the ground can be removed altogether with little observable difference.

While removing the ground plane may not make much sense on a circuit board, since it always has a ground plane, this situation is common in cabling. There, the "ground" can be an extra, grounded line on the outside of the cable called a "shield." The shield can be removed with little impairment. This is what Ethernet uses – unshielded twisted pair (UTP). This is important because, as one might guess, the unshielded cable is cheaper than shielded.

So are there situations in which differential propagation is less than ideal? There are two common cases, and their origins are unexpected.

Obstacles: Clock Jitter and Common-Mode Signals

One of the more subtle requirements for differential signaling to work correctly is that the two signals (V_1 and V_2) have to be synchronized in time. Backing up, there can be several reasons why the two are not in sync. For example, one line may be longer than the other line. There may be some asymmetry, such as a test pad or via. As we have noted, both of these cases are actually mistakes – recall that the line carrying V_1 needs to be as identical to the line carrying V_2 as possible. A more common cause of nonsynchronized signals lies inside the chip that creates the signals. Consider the clock signals inside the chip that drive the circuits that create V_1 and V_2 – if there is any difference in the clocks (a situation called *clock skew*), then the two signals will exit the chip at slightly different times.

So, consider what happens if the two outputs (V_1 and V_2) are out of sync. One of the outputs changes before the other, and so the differential voltage acquires a hiccup. This is illustrated below in Fig. 9.5.

Figure 9.5 illustrates what happens, but it needs some explanation.

In the top case, the two voltages are in sync. The common voltage is a DC voltage (halfway between V_1 and V_2) and the differential voltage is double the swing of a single signal.

In the bottom case, V_2 is delayed relative to V_1.

Consider the common voltage – V_1 rises first and then V_2 falls. The average of the two voltages first rises (one voltage is rising while the other stays the same) and then

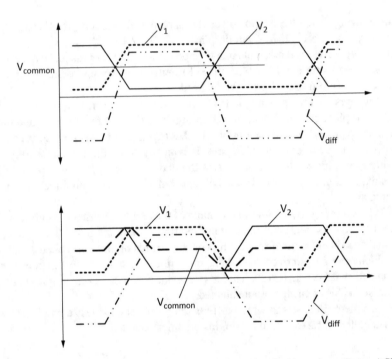

Fig. 9.5 Effects of clock skew on differential signals. (Top) Signal with no skew. (Bottom) Signal with V_2 slower than V_1

falls (the later voltage then falls). The common-mode voltage now has a small positive triangle-shaped "blip" in it, the width of which is roughly proportional to the skew (time difference) between the two differential voltages. Later, when V_1 falls, the "blip" is of opposite polarity because V_1 falls before V_2 rises.

Consider the differential voltage – When only one output switches, the differential voltage undergoes half of a voltage swing. Later, the other output switches and the remaining half-voltage-swing occurs. This creates a spread-out differential waveform, which looks like a slower-rise-time version of the original waveform. This is, in itself, not necessarily bad – it is the same signal but with a slightly slower rise and fall time.

So the result of a time delay between differential outputs is a train of impulses seen on the common-mode voltage and a slower rise and fall time seen on the differential voltage. The slower rise and fall time is likely negligible, but what does a train of impulses look like in the frequency domain? A train of impulses! The result is signal energy out to the clock frequency of the signal. A 1 Gbps SATA cable radiates out to 1 GHz.

And why is this a problem? First, the common-mode energy makes the cable act like a very large antenna. Second, there is no termination strategy for the common-mode signals. If you look back up to Fig. 9.3, it should be apparent that each of the lines, considered individually, is completely unterminated! So the impulses tend to run up and down the cable like a terrified squirrel at a wedding reception.

There are basically four ways to deal with this situation. The first is to eliminate asymmetries and skew. This is what is known in the industry as a root-cause fix – it eliminates the source of the problem. The second is to add back the shield. The shield can potentially help (e.g., it is required by the SATA standard) but it must be extremely well-grounded or else the shield will reradiate. (If it is not properly grounded, the shield turns back into a hunk of metal and therefore a big antenna.) The third option is to add a ferrite bead to the cable. The ferrite bead has the property that only signals that cancel themselves out (signals that are accompanied by a return path) do not "see" the inductance of the bead. So stray signals do "see" this inductance and see a much higher impedance. In other words, this train of pulses is attenuated by the inductance of the ferrite bead. This is discussed in more depth in the chapter on electromagnetic interference (EMI), but suffice to say it makes the cable cost more. Keep in mind that if you only pass EMI testing by using a cable with a ferrite, you have to ship a cable with a ferrite with every unit you sell. The unit is only legal for sale if it is accompanied with whatever special cabling was needed to make it pass the test.

The fourth way is to modify the termination so as to terminate common-mode voltage transients.

Termination Strategies Revisited

Consider a simulation of a differential signal with significant clock skew, shown in Fig. 9.6.

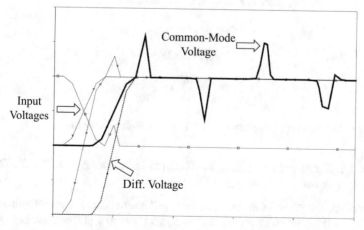

Fig. 9.6 Simulation of the case shown in Fig. 9.5

Fig. 9.7 Modified
termination strategy

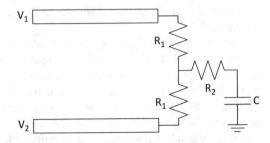

The simulation results are dramatic – the impulse train can clearly be seen. The results can be a little confusing. The input voltages shown in the figure are the two differential voltages V_1 and V_2. The differential voltage is $V_2 - V_1$, and is actually very well-behaved. In other words, this arrangement would function normally! The common-mode voltage is the voltage measured on one of the two differential lines, in this case V_2. This highlights what happens when each of the differential voltages is unterminated – a single transient results in a train of impulses due to reflections. Each of the single lines in the pair has the same train of impulses, and so the impulses subtract out when the differential voltage is calculated. However, they are present on the line and can cause radiated emissions.

So what is the fix? The solution is to add a common-mode termination path to the differential termination. In other words, one adds termination to each of the individual transmission lines. By adding a series resistor and capacitor at the halfway point, the differential signal still "sees" the same resistance, but each individual line sees half the differential resistance (i.e., it sees the correct single-signal terminating resistance) in series with the added resistor and capacitor. This is illustrated below in Fig. 9.7.

As shown in the figure, the differential termination has been split into two equal resistors labeled R_1 and the resistor R_2 and capacitor C have been added. For effective termination, there are some rules of thumb.

$$R_1 = 0.9Z_0 \tag{9.4}$$

$$R_2 = 0.1Z_0 \tag{9.5}$$

$$C = \frac{5t_r}{Z_0/2} = \frac{10t_r}{Z_0} \tag{9.6}$$

This if for the signal rise time t_r. If t_r is in ps, then C is in pF.

Example 9.1
Consider a differential pair with $Z_0 = 100\ \Omega$ (and therefore $Z_{\text{diff}} = 200\ \Omega$). The pair is driving a signal with a rise time of 800 ps.

(a) Calculate the resistors to terminate both differential and common-mode signals.
     $R_1 = 0.9Z_0 = 90\ \Omega$. $R_2 = 0.1Z_0 = 10\ \Omega$. $C = (10 * 800\ \text{ps})/100\ \Omega = 80\ pF..$
(b) Simulate the result.

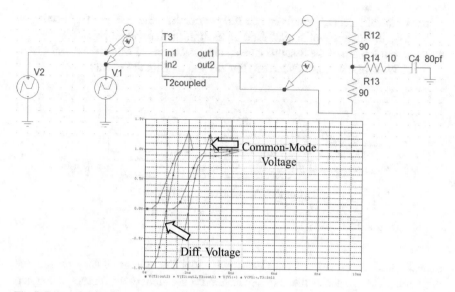

Fig. 9.8 Simulation of common-mode termination strategy

See Fig. 9.8 below.

As the simulation shows, the addition of the series resistor and capacitor has eliminated the train of impulses. (Contrast with Fig. 9.6.) There is only a single impulse because now the line is terminated, and the pulses do not reflect. However, there is still a single impulse, which shows that eliminating skew is still preferable to termination.

Appendix

Construct a SPICE simulation of two coupled dual transmission lines with LEN = 1, $L = 100$ nH, $C = 10$ pF, and LM = CM = 0. Connect the lines serially and connect a voltage source (800 ps rise time) at one end and a single terminating resistor across the outputs of the lines. (See the diagram below.)

1. Calculate the Z_0 of each line. For $R_S = 0$ (i.e., no source resistance) and $R_L = 2Z_0$, simulate the signal and show that the signal has "perfect" signal integrity.
2. Add a 2 nH inductor in series and a 5 pF shunt capacitor (to ground) to *one* of the signal lines at the midpoint between the two sets of lines. Which is affected more – the differential signal or the signal of a single line?
3. Restore the setup of the first problem (i.e., eliminate the L and C) and instead create 200 ps of clock skew between the two voltage sources. Do this by making

one of the differential signals switch 200 ps after the other one. What effect does it have on the differential signal? And on each of the individual signals?

4. Restore the setup of the first problem and instead change LM to 10 nH and CM to 5 pF. What effect does this have? How do you explain the result?

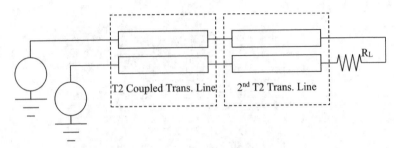

Example: Schematic of Problem 1 and Problem 4   Plot the voltage at each source, the differential voltage across the inputs (measured at the input to the first set of coupled transmission lines), and the differential voltage across the output (measured at the output of the second set of coupled transmission lines).

Consider a system that has a connector for a high-speed differential signal with a rise time of 200 ps. The system needs to terminate the signal correctly.

5. If the trace carrying the signal has an impedance of 75 Ω, what is the differential impedance needed for the signal? 150 Ω.

6. Calculate the R and C that needs to be added to the termination in order to terminate common-mode signals correctly. (See the example at the end of the lecture.) $R = 0.9 * 75 = 67.5\ \Omega$s. $R2 = 0.1 * 75 = 7.5\ \Omega$. $C = 10 * 200/75 = 26.7$ pF.

Chapter 10
Crosstalk

Background and Objectives

When this chapter is finished, you should be able to understand crosstalk. Crosstalk is, arguably, the best-known signal integrity problem, the archvillain of signal integrity. For that reason, it is commonly blamed for signal-integrity issues. Sometimes the blame is warranted, and sometimes not. It is important to understand what crosstalk is, so that you can identify when it is (and is not) causing your signal-integrity problems. After this chapter, you should be able to:

- Explain the physical origins of crosstalk
- Understand why near-end and far-end crosstalk are different
- Describe the circumstances under which crosstalk is eliminated
- Estimate, and then reduce or eliminate, crosstalk
- Understand how to recognize crosstalk in real-world situations

What Is Crosstalk?

Crosstalk occurs when one switching line injects noise into one or more neighboring signals. We know that mutual capacitance between lines goes up as lines get closer. (The lines behave like a parallel plate capacitor. Actually, the lines *are* a parallel plate capacitor.)

What about mutual inductance? As lines get closer, more magnetic-field lines overlap. Consider two signals with a center-to-center spacing of D spaced H above the ground plane, as illustrated in Fig. 10.1.

The mutual inductance, L_M, is proportional to

© Springer Nature Switzerland AG 2022
S. H. Russ, *Signal Integrity*, https://doi.org/10.1007/978-3-030-86927-4_10

Fig. 10.1 Crosstalk as a
function of spacing

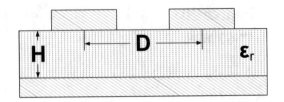

$$L_M \propto \frac{1}{1 + (D/H)^2} \qquad (10.1)$$

Example 10.1
What spacing reduces crosstalk by 90%?

$$L_M \propto \frac{1}{1 + (D/H)^2} = 0.1 \qquad (10.2)$$

$$\frac{1}{0.1} = 1 + (D/H)^2 = 10 \qquad (10.3)$$

$$(D/H)^2 = 9 \qquad (10.4)$$

$$D/H = 3 \qquad (10.5)$$

$$D = 3H \qquad (10.6)$$

This relationship is the source of the well-known *3H Rule*. The neighboring signals should be spaced at least 3H apart. For example, if signals are 0.2 mm above the ground plane, the signals should be spaced on a 0.6 mm center-to-center spacing.

This is what happens if signals are near each other and if there is a continuous, uninterrupted ground plane under both of them.

But what happens if there is a slot, or other discontinuity, in the ground plane underneath the signals? Consider two signals routed over a slot in the ground plane, as shown in Fig. 10.2.

The signal current is on the top of board, as shown in the two solid lines. The slot is shown in gray, and the return current (shown in dotted lines) must flow on the ground plane around the slot. The interruption of the ground plane forces the return current around the side. As you might imagine, by having the outgoing current diverted away from the return current, there is less cancellation of magnetic flux and therefore the self-inductance goes up. In addition, by forcing the return currents to share the same path back, the mutual inductance also increases dramatically.

So what causes crosstalk, exactly?

Fig. 10.2 A slot in the ground plane dramatically increases crosstalk

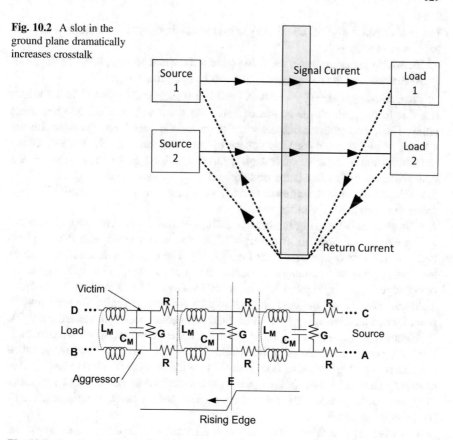

Fig. 10.3 Two lines with coupling between them

Near-End Versus Far-End Crosstalk

Consider two lines with mutual inductance and capacitance between them, as illustrated in Fig. 10.3.

In Fig. 10.3, line A-B is called the "aggressor"; it is the line carrying a changing signal flowing from source A to load B.

Line C-D is called the "victim." We are going to introduce special terminology to describe Line C-D for reasons that will become clear later. Point C on the victim line corresponds to the source of the aggressor line, and we call point C the "near end." So the near end is defined by the source side of the aggressor signal. Point D on the victim line corresponds to the load of the aggressor line, and we call point D the "far end." The far end is defined by the load side of the aggressor signal. (Technically both point B and point D are at the far end, but we are much more concerned about the victim signal in this chapter.)

Generally, C is the source of the victim signal and D is the victim's load because, usually, neighboring signals on a board travel in the same direction. This is definitely not always the case – C-D could carry some responding signal, in which case C is the load of the victim signal.

Returning to our example, the rising edge is shown at point E, between A and B.

At the moment the rising edge is at point E, two things happen.

First, $E_{\text{Victim}} = L_M \frac{dI_{\text{Aggressor}}}{dt}$. That is, there is a voltage pulse injected on the victim line due to the dI/dt on the aggressor line. One needs to remember two important points. First, the aggressor line has some characteristic impedance Z_0 and so there is a linear relationship between the voltage of the rising edge and the current. Stated differently, the rising edge of voltage is also a rising edge of current. Second, the voltage on the victim line is the first derivative of the current of the aggressor line. Since the current on the aggressor line is a ramp, the voltage on the victim line is a spike, the first derivative of the ramp.

This also leads to a very important question – does the current induced onto the victim line pass to the left or the right? The answer has to do with the coupling between the lines (better known as Lenz' Law). The current induced in the victim line tends to oppose the changing current in the aggressor line, and so if the aggressor rising edge is moving right to left, the victim voltage pulse is moving left to right. Thus the induced voltage is negative going to the far end (going forward toward point D) and positive going back to the near end (going backward toward point C).

What happens next is unexpected. As the rising edge moves from source to load, it creates a continuous stream of little pulses. Think of it this way – during the entire time the rising edge is moving from source to load (from A to B in this example), it is creating a continuous flow of little pulses on the victim line. Because of Lenz' Law, the pulses moving back to the near end are positive and the pulses moving forward to the far end are negative.

However, as the rising edge moves, the source of the little pulses move, as illustrated in Fig. 10.4. The rising edge is both continuously causing pulses on the victim line and continuously moving from source to load on the aggressor line.

At time T_1, the rising edge induces a positive backward-moving pulse toward the near end (L_1) and a negative forward-moving pulse toward the far end (S_1). At time T_2, the edge has moved down and introduces two new pulses, another positive backward-moving pulse (L_2) and another negative forward-moving pulse (S_2).

One thing to note is that the negative forward-moving pulses (S_1 and S_2) travel at the same speed as the rising edge. So the negative forward-moving pulse introduced into the line at time T_2 is superimposed on top of the pulse introduced at time T_1. In other words, the negative forward-moving pulses are going to add up.

A second thing to note is that the backward-moving pulses do not line up. The first pulse is emitted as soon as the rising edge enters the line, and the last pulse arrives at double the one-way propagation delay (one delay for the rising edge to propagate down the line and one more delay for the last pulse to get back).

The results are illustrated in Fig. 10.5.

The waveform at the victim *near end* is a long, small positive hump. The height of the hump is proportional to mutual inductance and inversely proportional to rise time. (That is, with smaller, faster rise times the crosstalk goes up.) The waveform at

Fig. 10.4 Induced current pulses as the rising edge propagates from source to load. Top: Rising edge is at one point at T_1. Bottom: Rising edge is at another point at T_2

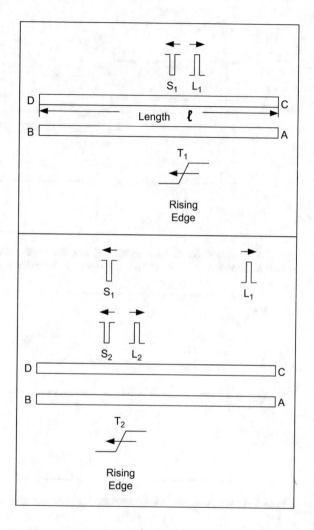

the victim *far end* is a tall negative spike. The height of the spike is not only proportional to mutual inductance and inversely proportional to rise time, but is also linearly proportional to the length of the lines. (More specifically, it is linearly proportional to the length over which the two lines are close together.)

The crosstalk at the victim's source is called *near-end crosstalk* or NEXT. The crosstalk at the victim's load is called *far-end crosstalk* or FEXT.

This is the unexpected result – the near-end crosstalk and the far-end crosstalk have completely different waveforms, even though they have the same physical origin.

It should be noted that, in Fig. 10.5, if the rising edge is replaced with a falling edge, all of the voltages invert. The reverse crosstalk is a negative hump and the forward crosstalk is a positive spike.

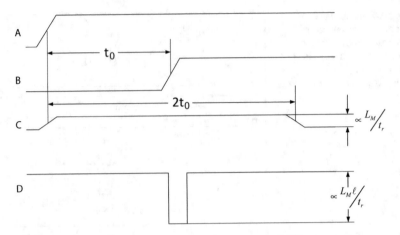

Fig. 10.5 Crosstalk waveforms at source and load due to mutual inductance. V_c is the crosstalk voltage at the victim's source and V_D is the voltage at the victim's load

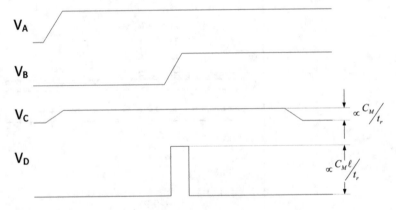

Fig. 10.6 Crosstalk waveforms at source and load due to mutual capacitance

Second, $I_{\text{Victim}} = C_M \frac{dV_{\text{Aggressor}}}{dt}$. This pulse, unlike the one due to inductance, is a current pulse and has the same polarity as the voltage. Since it is an impulse of current, it creates a positive voltage going both back to the near end *and* forward to the far end. Remember that these are instantaneous pulses, and so the ratio of voltage to current is constant, namely Z_0. So a current pulse on the victim line turns into a voltage pulse. The results are shown in Fig. 10.6.

This looks similar to Fig. 10.6, except the negative spike is now a positive spike. In other words, the NEXT looks the same but the FEXT is of opposite polarity.

Third, the total crosstalk is the sum of the inductive and capacitive crosstalk. In the reverse direction, the NEXT simply adds up and the result is a long positive hump that lasts twice the propagation delay. In the forward direction, the FEXT actually tends to cancel out since the inductive spike is negative and the capacitive spike is positive.

It turns out that if the dielectric surrounding the two signals is homogenous (meaning "all the same" or "uniform"), the two crosstalks are exactly equal and so the FEXT exactly cancels to zero. This is the actual case in stripline, when signals are routed between power and ground planes. The signals are completely surrounded by circuit-board material.

So what if they are non-homogeneous? For example, what about microstrip, in which the signals are on the top layer of a circuit board? The dielectric surrounding the conductors is half air, half FR-4 fiberglass (or whatever the board is made of). So the relative dielectric constant drops and there is less crosstalk due to mutual capacitance. So forward crosstalk is inductive, a muted negative spike, less than the spike caused by mutual inductance but significantly greater than zero.

So we have seen that crosstalk introduces signals shaped like spikes and humps onto neighboring, quiet lines. If the lines are unterminated, the crosstalk will reflect and create a large mess. If lines are terminated, the crosstalk makes a one-way trip and then stops. There is still crosstalk, but at least it only occurs once.

Estimating and Reducing Crosstalk

To summarize what we have learned so far, the amplitude of near-end crosstalk is proportional to the sum of mutual inductance and capacitance and inversely proportional to rise time. The time duration of the "hump" is proportional to the coupled length of the two lines (i.e., the length over which the two lines are tightly coupled). Far-end crosstalk is proportional to the difference of mutual inductance and capacitance, inversely proportional to rise time, and linearly proportional to the coupled length of the two lines. So, in both cases, as rise time goes down (gets faster), the crosstalk is worse. As the coupled length increases, far-end crosstalk increases in amplitude and near-end crosstalk increases in time duration.

The steps to reduce crosstalk arise naturally from this information. Increasing the rise time (i.e., slowing down the signal) will always reduce crosstalk, but may not be an option if the signal is constrained. (For example, SATA has minimum rise time constraints.) If the rise time can be made slower, a series resistor at the source will usually do the trick; it creates an RC filter (using the capacitance of the wire) that increases the rise time.

Reducing mutual inductance and capacitance will help. This generally involves increasing the spacing between the lines. This also reduces the coupled length of the two lines.

One solution for crosstalk that is often proposed in books and articles is to ad *guard traces*. Guard traces are grounded wires placed in between signals. Sometimes the picture is painted that the ground of the guard trace somehow magically sucks the crosstalk out, but this is not the case. So the answer to the question "do guard traces help?" is actually "yes, no, and yes."

First, if you insert a guard trace between two traces with crosstalk, the crosstalk will indeed go down. So the first answer is "yes."

Second, it turns out that the reduction is actually mostly due to the fact that you had to spread the signals apart to insert the guard traces. You could have accomplished the same reduction without the guard trace just by moving the signals apart. So the second answer is "no."

Third, and I only learned this myself when I started managing staff doing circuit-board layout, guard traces force physical spacing, and give the circuit-board designer a tangible way to make sure the spacing is adequate. Since the guard trace is harmless, it is an effective way to enforce extra spacing rules between signals likely to have crosstalk issues. So the third answer is "yes."

So how do you know you have a crosstalk problem?

Diagnosing Crosstalk

Crosstalk usually takes place where different buses are routed near each other, say SATA and PCIe. Signals that share a bus typically have few crosstalk issues because signals on a bus all tend to switch at the same time. (This is roughly analogous to saying that people yelling at a ball game are rarely disturbed by their neighbors' yelling.) The crosstalk runs around the circuit traces during the time all the edges are rising and falling anyway. Conversely, when two different buses are near each other, all of the signals changing on one bus can "gang up" on the other bus nearby. Modern memory buses are really bad (from a crosstalk standpoint) because they are fast, constantly changing, and typically involved dozens of bits.

So how do you discover a crosstalk problem?

Engineer's Notebook: Crosstalk Problems

We had designed our set-top box around a new type of plug-in card. The card was a beast, 68-pin connector and dozens of switching signals. We routed in on the board next to our SATA bus, mainly because the two buses came out next to each other on the rear panel. (Notice how the rear-panel design tends to drive your board layout which tends to drive signal integrity.)

The software for the plug-in card was late, so we were well into our prototyping when the software team started using it. They reported that the hard drive stopped working every time they used the plug-in card.

I need to stop here and point out that there are really only two ways a switching bus can ruin another bus, ground bounce and crosstalk. These are therefore always the "usual suspects" that you go look for.

To help decide which it was, we got out the circuit-board layout and looked at it. Sure enough, the SATA bus ran alongside the card bus for several inches and both were right next to each other. At this point, the culprit is almost certainly crosstalk.

One might suspect ground bounce still, but no other signals seemed to have ground bounce. Recall that ground bounce tends to come squirting out of quiet chip outputs. Other signals would have almost certainly manifested ground bounce earlier. (If we had wanted to, we could have probed quiet signals and look for the famous double-pulse of ground bounce.)

So we changed the board layout by adding a fat guard trace between the SATA and the card bus, and, sure enough, the crosstalk went away. (Why a fat trace? It basically boiled down to "why not?")

The lesson of the story is this. Crosstalk may be difficult to spot on an oscilloscope, but it is easy to diagnose with the symptoms. When turning on one bus causes another bus to misbehave, this is usually a sign of crosstalk. This also shows the importance of testing with every part of the system running simultaneously.

Appendix

Construct a coupled triple transmission line with LEN $= 5$, $L = 100\text{nH}$, $C = 10\text{ pF}$, LM $= 10\text{nH}$, and CM $= 5\text{ pF}$. Connect a source to input 1 and 2 and a load to input 3 (and the opposites to outputs 1–3). See the diagram below, and note which sides have the source and which have the load. The "source" in lines 1 and 3 is just the source impedance – no voltage source – and input 2 has a voltage source in series with its source resistance and a 500 ps rise time.

1. Compute the Z_0 and T_p of the transmission lines. (Note that the Spice L is L_L and the Spice C is C_L.)
2. Simulate with no termination ($R_L = 1\text{ M}\Omega$ and $R_S = 10\text{ }\Omega$). Note where R_S and R_L are located!
3. Repeat the simulation, but with load termination ($R_L = Z_0$ and $R_S = 10\text{ }\Omega$).

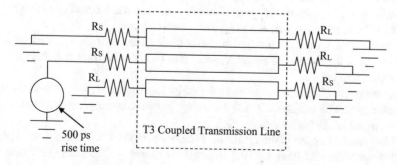

Example: Schematic of Problems 2 and 3   For both simulations, plot the voltage at the source and load of the second line and the loads of the first and third lines.

Three lines are routed in parallel, carrying signal X, signal Y, and signal Z. All three lines are 50-Ω transmission lines that are routed very close together, and are all routed microstrip, on the top layer of a circuit board directly over a ground plane, with an insulation thickness of 0.7 mm. The lines run parallel to each other and remain closely spaced over a length of 6 inches. The source of signal Z is on the same side as the source of signal Y (and therefore the loads are on the same side as well). Signal X runs in the opposite direction, with its load at the source of Y, and its source at the load of Y. See below for a diagram.

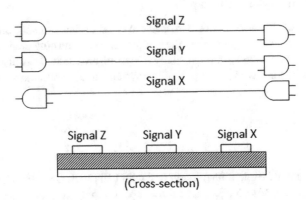

(Cross-section)

Signal Y has a rising edge with a fast (100 ps) rise time. Signal X and Z are both "quiet" when signal Y switches.

To reiterate an important point: Crosstalk is considered both at the "near end" and the "far end." The near and far end are from the point of view of the *aggressor* signal. Whether the victim signal's logic input is at the "near end" or "far end" depends on the victim signal. Continuing on with the problem. . .

4. Since the line is routed microstrip, do you expect the FEXT to be inductive, cancelled out, or capacitive? Microstrip = non-uniform dielectric = not cancelled out. Air has lower capacitance, so *inductive FEXT* will dominate.

5. When signal Y rises, what signal is seen at the logic input of Signal X (the point where the wire enters the input of a logic gate, not the point where the wire exits the output of the other logic gate)? Since the X input is at the near end of signal Y, NEXT waveform, which is a long, low, positive hump.

6. When signal Y rises, what signal is seen at the logic input of Signal Z? Since the Z input is at the far end of signal Y, inductive FEXT waveform, which is a short negative spike.

7. Assuming the lines are still routed microstrip, identify three changes that could be made to reduce crosstalk. Increase spacing, slow the rise time, move lines closer to ground plane (thinner dielectric).

8. What would happen to the crosstalk if someone added a slot to the ground plane that ran across the three signals? It would get worse – more inductive coupling.

Alice was concerned about crosstalk, and so she instructed Bob to re-route the signal using stripline instead of microstrip.

9. So where are signals X, Y, and Z found after the change? (Think vertically, and assume the horizontal placement is unchanged.) Where are the signals found vertically after the change? Between the power and ground planes.

10. Since the line is routed stripline, do you expect the FEXT to be inductive, cancelled out, or capacitive? Homogenous dielectric = > no FEXT (cancels out).

11. When signal Y rises, what signal is seen at the logic input of Signal X (the point where the wire enters the input of a logic gate, not the point where the wire exits the output of the other logic gate)? Still see a NEXT waveform.

12. When signal B rises, what signal is seen at the logic input of Signal Z? No crosstalk is observed.

Chapter 11
Power Distribution Network: Frequency Domain Analysis

Background and Objectives

We have studied "lumped" parameters such as resistance, capacitance, and inductance and saw that the combination of inductance and capacitance causes a power-related effect called ground bounce. This chapter studies the behavior of the power supply. As we will see, it serves both a DC and an AC purpose, and so stands astride the two worlds. When this chapter is finished, you should be able to:

- Describe the demands on a power distribution network
- Bound the frequencies over which a power distribution network must operate
- Describe the behavior of a power distribution network in the frequency domain
- Understand the resonances that are present in a power distribution network
- Design a power distribution network using a combination of calculations and simulation
- Select bypass capacitors and understand how to add them to a design

The Power Distribution Network

So, what is the role of the power distribution network on a circuit board? Clearly, it is to distribute power. But what is it really used for? Real power? Imaginary power?

The first assumption everyone makes is that, since the voltages to almost all circuit-board components are DC voltages, the power supply is a DC element. It turns out that this is only partially true. In fact, most digital systems (specifically, systems built using CMOS technology) consume almost no DC power at all. If the clock to a CMOS chip is frozen, the power consumption drops dramatically. (Some CMOS circuits are dynamic, meaning that if the clock is shut off they do not work correctly, but that is an implementation detail.)

© Springer Nature Switzerland AG 2022
S. H. Russ, *Signal Integrity*, https://doi.org/10.1007/978-3-030-86927-4_11

Most of the power consumption is the result of AC activity. Unfortunately, when we all learn electrical engineering, we tend to think of AC activity in terms of sine waves. AC power is a sine wave of voltage, for example. But if you go back to the earliest classes, AC can refer to any sort of voltage or current that is not DC. In the context of power supplies, what matters is the transient response. To make the analysis easier, we will treat transients as AC phenomena and will use the rise time/ knee frequency formula to convert a transient into its frequency-domain counterpart.

Inside a digital integrated circuit, most of the power consumption is due to switching activity, of changes to internal signals, and driving capacitive loads. (This is the $\frac{1}{2}CV_{cc}^2$ formula discussed earlier.) Some power consumption may be the result of output activity – like activating a radio transmitter or video output.

At any rate, the power is consumed in nearly instantaneous impulses (or perhaps more accurately, transients), not in a smooth waveform. See, for example, Fig. 5.1 (in Chap. 5), which shows the current flow (I_C) during a signal-change event. The power drawn from the power distribution network is ultimately the sum of billions of these waveforms.

The result of this is that the power distribution network has to supply both DC power and AC power, and it has to supply AC power up to very high frequencies. In fact, as you have probably already guessed, it has to supply power all the way up to the knee frequency of each signal.

The ideal power supply has zero resistance and, because of the AC requirement, zero impedance. So the requirement of a power supply, to state it formally, is this: A power supply has to have low impedance from DC up to the highest knee frequency in the system being powered.

This raises four important questions:

First, how low is "low" impedance? The exact answer is "it depends," as in "it depends on the needs of the circuits you are designing." It can be calculated or estimated, as will be seen, and sometimes it is specified on the datasheet for a large integrated circuit. (In a complete vacuum of knowledge, 0.1 Ω is a good guess.)

Second, what keeps power supplies from having low impedance at high frequency? The short answer is inductance. Every circuit element is inductive and so the power-supply impedance will inevitably reach a point in the frequency domain at which inductance dominates.

Third, what happens when it is wrong, when the impedance is too high? When an integrated circuit tries to pull current out of the power supply faster than the inductance of the power-supply wiring permits, it creates ground bounce. The result is that the difference between the power voltage and ground drops. Whether this is due to the ground going up or the power rail going down does not matter from a functional standpoint – the apparent power-supply voltage sags or droops.

If the droop is large enough, the circuits inside an integrated circuit stop working correctly. Memory cells, such as SRAM and D-flip-flops, can lose their state. Logic gates do not work correctly. So the amount of droop allowable at the input of an integrated circuit is a function of the IC's noise margin, and can generally be read from the datasheet of the IC.

One subtle aspect of the issue is that it may only manifest when the circuit runs at maximum speed and maximum data-processing intensity. This is why power supplies really need to be tested with processors executing software, with all peripherals turned on, and with frequent access to memory.

Fourth, what can an engineer do about it? An engineer can add the one component that has *lower* impedance at *high* frequency – capacitors. These capacitors are called *bypass capacitors* because they serve to bypass temporarily the main power supply.

The role of bypass capacitors is to function as a current bank. (Here the term bank is being used in the sense of a metaphorical financial institution.) If you recall the discussion on capacitors, they serve as voltage-controlled buckets of charge. If the voltage drops, they dump out the charge. So they can source current if and when the power-supply voltage drops. They source enough current to keep the integrated circuits working (i.e., to keep the power-supply voltage high enough) until the inductive power supply can "catch up" and source current.

What makes the process interesting is that the capacitors themselves have inductance, and so an actual design process is needed.

But before learning the process, one needs to study resonance. . .

Power Supplies and Resonance

As one considers a power supply in the frequency domain, there are places where the inductances and capacitances interact in unexpected ways.

But first, a review. . .

Consider an inductor. As students learn early on, it has an impedance equal to $j\omega L = j2\pi fL$. If one were to graph the magnitude of impedance versus frequency, it would be linear and the slope would be proportional to the inductance.

Consider a capacitor. Its impedance is $1/j\omega C$, and so its graph of the magnitude of its impedance looks like a reciprocal curve.

But what if the two curves were plotted on a log-log scale (log of impedance versus log of frequency)?

Consider the inductance. First, the logarithm of the inductance turns out to be the logarithm of the frequency plus the logarithm of $2\pi L$, which is a constant. So the curve of the logarithm of the impedance is a logarithm curve that is moved vertically up and down depending on the value of inductance. Second, when plotted on a logarithmic horizontal axis, the logarithm curve "straightens out" to a line of slope + 1. So the result is that the impedance of an inductor, when plotted on a log-log scale, is a line of slope + 1 that "slides around" depending on the inductance. If the inductance drops, you can think of it as dropping vertically or *sliding to the left*. It is symmetric (i.e., you can think of it either way) because it is a line of slope + 1. This is illustrated in Fig. 11.1.

Everything also holds true for a capacitor, as is shown in Fig. 11.1, but with two differences. First, the "height" is a function of $1/C$, so increasing capacitance moves the line to the left, not to the right. Second, the slope is −1 instead of +1.

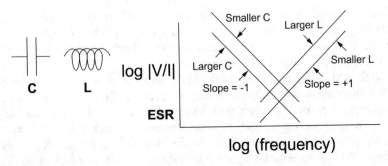

Fig. 11.1 The impedance of a capacitor and inductor on a log-log scale. Changing the value of C or L "slides" the line horizontally in frequency

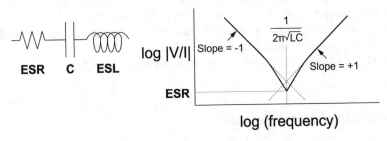

Fig. 11.2 Equivalent circuit of a capacitor, and its impedance in the frequency domain

Armed with the knowledge of how impedances of passive elements appear on a log-log scale, one can now consider two special cases of inductance and capacitance. These special cases are called *resonances*.

The graph in Fig. 11.1 bears some review. It is a graph of impedance (y-axis) versus frequency (x-axis) plotted on a log-log scale. As a result, the graph of an inductor has a slope of +1 and a change in the value inductance moves the line vertically. The graph of a capacitor has a slope of −1 and, as with the inductor, changing the capacitor's value (i.e., changing the amount of capacitance) moves the line vertically. We will use this type of graph throughout this chapter.

For the first resonance, consider a capacitor by itself. It has capacitance and, as we have seen, has ESR and ESL. This is illustrated in Fig. 11.2.

This type of resonance is called a *series RLC resonance*. In the region below f_{res}, the capacitance dominates and the impedance drops linearly with frequency. In the region above f_{res}, the inductance dominates and the impedance rises linearly with frequency. At f_{res}, the impedances of the inductor and capacitor exactly cancel and the impedance is at a minimum, purely real, and equal to the ESR.

As you may recall from our discussion of a capacitor's self-resonant frequency (SRF), we have

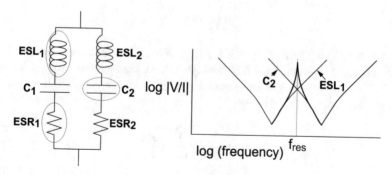

Fig. 11.3 Equivalent circuit of two different capacitors

$$f_{res} = \frac{1}{2\pi\sqrt{LC}} \qquad (11.1)$$

where L is the ESL and C is the value of the capacitor. At f_{res}, $Z = $ ESR. Not to put too fine a point on it, below f_{res}, the impedance curve has a slope of -1 and, above f_{res}, it has a slope of $+1$.

The second resonance involves pairs of capacitances. As will be seen later, larger-valued capacitors have more inductance. (They are physically larger and are usually placed into inductive packaging for cost reasons.) The large capacitance value is necessary – switching circuits can demand hundreds of Amps for a few nanoseconds – with the result that the capacitance value has a (relatively) high inductance. Thus the larger-valued capacitors have a limited frequency range, and so must be combined with lower-inductance, smaller-valued capacitors. The second resonance, then, involves the parallel combination of two capacitances.

So consider a "real" capacitor (one with C, ESR, and ESL) in parallel with a different "real" capacitor. The first capacitor is a large-valued capacitor with relatively high inductance and the second is a smaller-valued capacitor with relatively low inductance. Since the inductances and capacitances are different, there is a frequency region in which the larger capacitor's operation is dominated by ESL and the smaller capacitor's operation is dominated by capacitance. (Thus, this frequency region of interest is the region above the first capacitor's SRF and below the second capacitor's SRF.)

This is illustrated in Fig. 11.3.

This arrangement can have a type of resonance called a *parallel C-series RL resonance*. In the region below f_{res}, the net impedance is dominated by the ESL of C_1 and, above f_{res}, by the capacitance of C_2. (This is very confusing, but keep in mind that this f_{res} is different from the previous f_{res} because this is a different type of resonance.)

The derivation for this resonance is much more complicated, but it boils down to this. First, whether the resonance exists at all depends on a calculation. Specifically, this resonance only exists if

$$L/C > R^2 \tag{11.2}$$

Note that L is the ESL of C_1, R is the ESR of R_1, and C is the capacitance of C_2. (In other words, it is the R and L of the high-valued, high-inductance capacitor and the C of the low-valued, low-inductance capacitor. Very confusing!) If it does exist, then the frequency is found using

$$f_{res} = \frac{1}{2\pi}\sqrt{\frac{1}{LC} - \frac{R^2}{L^2}} \tag{11.3}$$

At f_{res}, the impedance is purely real and at a maximum (if the resonance exists). The impedance is calculated

$$Z_{res} = L/RC \tag{11.4}$$

This impedance is potentially more serious because Z_{res} is a *maximum*. That is, since it is a maximum, it creates a situation that is more likely to exceed the desired maximum impedance.

Armed with the knowledge of the log-log impedance graph and of resonances, we can proceed.

Design Strategy: The Big Picture

In a typical design, the power supply is on a separate circuit board, and the DC power arrives over some cabling. This offers very high inductance, so a few large capacitors are added at the connector where the cable arrives on the board. Most engineers describe these capacitors as "filtering" the incoming power, but in reality, they are sourcing current when the instantaneous power demands are too high. They are filtering the transients by supplying transient power. In this discussion, we will call this capacitor the "filter capacitor," although there is no official name for it.

These capacitors are only effective up to tens of Megahertz, leaving a substantial gap between the capacitors and the knee frequency. In order to span all the way up to the knee frequency, an extremely small inductance is needed. The "trick" is to use dozens of small capacitors in parallel, which lowers the effective inductance. It is these smaller capacitors that are typically called "bypass capacitors." By putting a few around every integrated circuit on the board, each IC "sees" a low-inductance power supply.

The design process entails selecting the right size of bypass capacitor and the right number of bypass capacitors, and finishes with simulation to confirm that it will work.

The design strategy will systematically move up the frequency spectrum from DC to the highest knee frequency of the system. The strategy starts with an inductance and finds the point at which the impedance of the inductance becomes too high. It then adds a capacitance that keeps the impedance low. Since real capacitors have inductance, the added capacitor eventually becomes an inductor and the process repeats.

One crucial property is this: If two impedances are parallel, the net impedance is roughly the smaller of the two. For example, a 2-Ω resistor in parallel with a 100-Ω resistor presents roughly 2 Ω of resistance. So if a capacitor is placed in parallel with an inductor, the net impedance is the minimum of impedance of the inductor and the impedance of the capacitor. Thinking back to the graph in Fig. 11.3, this means that when two capacitors (of different values) are placed parallelly, the net impedance versus frequency is the minimum of the two, and so think of the curve as tracing out the bottom of every hump.

Design Strategy: The Details

Step 1: Estimate the allowable power-supply impedance

Sometimes a chip manufacturer will call out a maximum tolerable power-supply impedance. (Intel routinely does this, for example.)

If no such estimate is available, one can estimate the current needed. First, since $I=CdV/dt$, the current for each output is the load capacitance, times the amount of voltage of a logic swing, divided by the rise time. Second, estimate the tolerable voltage droop, typically 2–10% of a full logic swing. Third, the estimated maximum power-supply impedance is the voltage droop divided by the current.

If all else fails, guess 0.1 Ω.

Regardless of the calculation method, the maximum tolerable power-supply impedance is called X_{max}.

Step 2: Size the big filter capacitor

First, estimate the inductance and resistance of the power-supply wiring, L_{psw} and R_{psw}. The round-conductor inductor formula will typically work.

Second, the power-supply wiring will have too much impedance at the point where the impedance of the inductance exceeds the tolerable power-supply impedance. Stated differently, the impedance of the power-supply wiring is primarily inductive, and so it increases linearly with frequency. At some point, the magnitude of the impedance will exceed the maximum allowable power-supply impedance. This frequency is called f_{psw} and is found by

$$f_{psw} = \frac{X_{max}}{2\pi L_{psw}} \tag{11.5}$$

(In other words, solve for the frequency at which $\omega L = X_{max}$.)

Third, size the filter capacitor to have an impedance below X_{max} at f_{psw}. The capacitor size is C_{filter}.

This may seem a little complicated but it is not. Impedances in parallel have the property that the total impedance is always less than the smaller of the two. If there is an inductor in parallel with the capacitor, the impedance of the inductor is going up with frequency and the impedance of the capacitor is going down with frequency. At extremely low frequency, the inductor has lower impedance and so the impedance of the parallel combination is more or less that of the inductor. Conversely, at high frequency, the impedance is that of the capacitor. As the frequency goes up, the impedance rises up to a point at which the two impedances are equal, and then will start to go back down again. Refer back to Fig. 11.3 for a diagram of this type of behavior.

So the filter capacitor is selected so that its impedance is X_{max} at the point where the impedance due to L_{psw} reaches X_{max}. In other words, the filter capacitor starts to take over and keep the impedance low at the point (in the frequency domain) where the power-supply wiring becomes unusable. By solving $X_{max} = 1/j\omega C$, we have

$$C_{filter} = \frac{1}{2\pi X_{max} f_{psw}} \tag{11.6}$$

Fourth, find the ESL of the filter capacitor. This might be in the data sheet, but can always be inferred from the self-resonant frequency. (Also note the ESR of the capacitor, R_{filter}.)

$$L_{filter} = \frac{1}{C(2\pi f_{SRF})^2} \tag{11.7}$$

Fifth, find the point where the filter capacitor is no longer effective due to the ESL of the bypass capacitor, L_{filter}. This calculation is identical to the one used for L_{psw}. This frequency is called f_{filter}.

$$f_{filter} = \frac{X_{max}}{2\pi L_{filter}} \tag{11.8}$$

Step 3: Size the array of bypass capacitors keeping below X_{max} from f_{knee} down to f_{filter}

This is more complicated than the filter capacitor calculation because, first, it is bounded on both sides of the frequency domain and, second, it involves adding an array of capacitors instead of a single capacitor. That is, we are now going to add an entire array of bypass capacitors. An array of small capacitors in parallel is needed in order to keep the effective inductance low.

First, at the high end of the frequency domain, we know $f_{knee} = 1/2t_r$. In other words, f_{knee} is the upper bound on the frequencies of the power-supply transients.

Using the inductance formula and using $2\pi f_{knee} = \pi/t_r$, we can use this as an upper bound on the inductance of the bypass capacitor array, L_{array}.

$$L_{array} = \frac{X_{max} t_r}{\pi} \tag{11.9}$$

To be clear, L_{array} is the net inductance of all of the bypass capacitors in parallel.

Second, find the ESL of the bypass capacitors in the array. As with the ESL of the bypass capacitors, this parameter might be in the data sheet, but can always be inferred from the self-resonant frequency. (Also note the ESR of the capacitor you intend to use, R_{cap}.)

$$L_{cap} = \frac{1}{C(2\pi f_{SRF})^2} \tag{11.10}$$

(Keep in mind that the subscript "cap" is the value for a single bypass capacitor and the subscript "array" is the value for the entire array of bypass capacitors.) If the self-resonant frequency is not on the datasheet, consult online sources to find qualified estimates of the inductance given the size of the package you plan to use. Surface-mount packages typically have a certain inductance and, conversely, the inductance of a bypass capacitor is typically a function of the package that is chosen.

Third, select the minimum number of capacitors needed in parallel to bring the inductance of the array to a sufficiently small value. This is the "trick": By using enough capacitors in parallel, the effective inductance of the array of capacitors is small enough.

$$N_{array} \geq \frac{L_{cap}}{L_{array}} \tag{11.11}$$

Fourth, the total capacitance of the entire array must be sufficiently large so that it is below X_{max} at f_{filter}. That is, the impedance of the array has to start being low enough to "take over" at the point in the frequency domain where the filter capacitor is no longer effective. This calculation is very similar to the calculation used to size the bypass capacitor.

$$C_{array} = \frac{1}{2\pi X_{max} f_{filter}} \tag{11.12}$$

The capacitance value of a single bypass capacitor is the array capacitance divided by the number of capacitors.

$$C_{cap} = \frac{C_{array}}{N_{array}} \tag{11.13}$$

The result is the minimum capacitance value C_{cap} and the minimum number of capacitors N_{array}. Increasing C_{cap} (e.g., to a very common capacitor value like 0.1 μF) will provide additional margin. Increasing N_{array} will also help, but often is not practical.

Step 4: Find the capacitance of the power and ground plane

For plane size A (in^2) and plane separation d (in), the capacitance of the power plane in pF, C_{plane}, is simply

$$C_{plane} = \frac{0.225\varepsilon_r A}{d} \qquad (11.15)$$

or, for the special case of FR-4,

$$C_{plane} = \frac{0.944A}{d} \qquad (11.16)$$

These formulas are derived by using the parallel-plate capacitor formula and factoring in all of the unit conversions. The power and ground plane effectively have 0 resistance and 0 inductance.

Step 5: Calculate the resonances

If you want to calculate the resonances, it is not hard as long as you carefully keep track of which R, L, and C to use.

1. To find the resonance between power-supply wiring and the filter capacitor:

   (a) Use R_{psw} and L_{psw} of power-supply wiring and C_{bypass} of bypass capacitor.
   (b) See if Parallel C – Series RL resonance exists.
   (c) If so, compute f_{res} and Z_{res}.
   (d) If $Z_{res} > X_{max}$, redesign the bypass capacitor (e.g., larger capacitance).

2. To find the resonance between the filter capacitor and the array of capacitors:

   (a) Use R_{bypass} and L_{bypass} of bypass capacitor and C_{array} of capacitor array.
   (b) See if Parallel C – Series RL resonance exists.
   (c) If so, compute f_{res} and Z_{res}.
   (d) If $Z_{res} > X_{max}$, redesign the capacitor array (e.g., larger capacitance).

3. To find the resonance between the array of capacitors and the power/ground planes (the most likely one of the three to exist):

   (a) Use R_{array} and L_{array} of the capacitor array (R_{cap}/N_{array} and L_{cap}/N_{array}) and C_{plane} of the planes.
   (b) See if Parallel C – Series RL resonance exists.
   (c) If so, compute f_{res} and Z_{res}.
   (d) If $Z_{res} > X_{max}$, redesign the capacitor array (e.g., more capacitors to lower R and L).

Design Strategy: The Role of Simulation

Once all the parameters have been calculated, the entire circuit can easily be simulated in SPICE or some similar software package.

A SPICE simulation can be run in AC or DC; the obvious choice here is AC. The sweep needs to run from below L_{psw} to past the knee frequency.

The result of the simulation needs to be a measure of impedance in ohms. When one considers the eq. $V=IR$, if the current is 1 Amp, then $V = R$. So, if the circuit is driven by an AC current source of 1 Amp, the voltage is equal to the impedance. The 1-Amp AC current source can then be swept in frequency.

This leads to a very confusing simulation – what is the physical significance of a 1 Amp AC current source being swept from DC to the knee frequency? The best way to think of it is this: The AC source is actually simulating switching transients on the power-supply network (PSN). At AC, the DC power supply has an impedance of zero, and so the power supply itself vanishes from the simulation – it turns into a wire to ground.

Most of the parameters are straightforward. The one that can be confusing is the array of capacitors. When running the simulation, use C_{array} for capacitance, ESR/N_{array} for resistance, and L_{cap}/N_{array} for inductance.

Example 11.1

Design a power distribution network for a simple design. The design has a target X_{max} of 0.1 Ω, a rise time of 500 ps, and a power-and-ground-plane capacitance of 1500 pF. The power-supply inductance is 175 nH. The ESL of the filter capacitor is 8 nH and its ESR is 0.05 Ω. The ESL of a single bypass capacitor is 2 nH, and the ESR of a single bypass capacitor is 0.1 Ω.

(a) Compute f_{psw}, the point at which the power-supply wiring has too much impedance.

$$f_{psw} = X_{max}/2\pi L_{psw} = 0.1/(2 * \pi * 175 \text{ nH}) = 90.95 \text{ kHz}$$

(b) Compute C_{filter}, the amount of filter capacitance needed.

$$C_{filter} = 1/2\pi X_{max} f_{psw} = 1/(2 * \pi * 0.1 * 90.95 \text{ kHz}) = 17.5 \text{ μF}$$

(c) Compute f_{filter}, the point at which the filter capacitor has too much impedance.

$$f_{filter} = X_{max}/2\pi L_{filter} = 0.1/(2 * \pi * 8 \text{ nH}) = 1.99 \text{ MHz}$$

(d) Compute L_{array}, the maximum permissible inductance of the bypass capacitor array.

$$L_{array} = X_{max} t_r / \pi = 0.1 * 500 \, ps / \pi = 15.9 \, pH$$

(e) Compute N_{array}, the number of array capacitors needed.

$$N_{array} = L_{cap} / L_{array} = 2 \, nH / 15.9 \, pH = 125.66$$

(f) Compute C_{cap}, the capacitance value of each array capacitor.

$$C_{array} = 1/2\pi \, X_{max} f_{byp} = 1/(2 * \pi * 0.1 * 1.99 \, MHz) = 0.800 \, \mu F$$

$$C_{cap} = C_{array} / N_{array} = 0.800 \, \mu F / 125.6 = 6.366 \, nF$$

In practice, C_{cap} would probably be rounded up to 10 nF (also known as 0.01 μF)

(g) Compute R_{array}, the resistance of the entire capacitor array, taking into account that there are N_{array} capacitors in parallel.

$$R_{array} = 0.1 / 125.66 = 0.7958 \, m\Omega$$

(h) For the resonance between the capacitor array and the power and ground planes, compute the frequency of resonance and the impedance at the resonance.

Does the resonance exist?

$L_{array} / C_{plane} = 15.9 \, pH / 1500 \, pF = 0.0106$ and $R_{array}^2 = 0.7958$ $m\Omega = 0.000000633$. Since $L/C > R^2$, the resonance exists.

$$f_{res} = \frac{1}{2\pi} \sqrt{\frac{1}{L_{array} C_{plane}} - \frac{R_{array}^2}{L_{array}^2}} = 1.030 \, GHz$$

$Z_{res} = L_{array} / R_{array} C_{cap} = 13.33 \, \Omega$. This is over 130 times larger than the maximum allowable power-supply impedance!

(i) Is the resonance a problem?

The bandlimit of the system is $1/2t_r = 1/(2*500 \, ps) = 1 \, GHz$. The resonance is very close to the knee frequency (with 3%) and so this is a marginal design. Worse, the resonance involves the entire power and ground planes and so will radiate very effectively.

(j) Simulate the resulting circuit in SPICE and compare with the calculations.

The SPICE schematic is shown in Fig. 11.4.

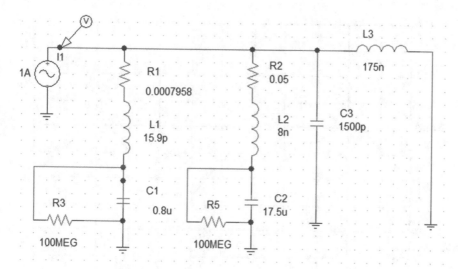

Fig. 11.4 Schematic of power-supply network

In the schematic, R1, L1, and C1 are the bypass capacitor array, R2, L2, and C2 are the filter capacitor, C3 is the capacitance of the planes, and L3 is the power-supply wiring. (The 100MEG resistors are added for SPICE to do the DC calculation correctly.) As predicted, the power supply has vanished from the simulation – specifically, it is the wire that runs from L3 to ground.

Again, the 1A current source is a fictional construct that permits one to graph impedance versus frequency. By using a value of 1 Amp, the voltage displayed on the SPICE simulation is equal to the impedance in ohms. You can also think of the current source as the point of view of an integrated circuit on the board with a 500 ps rise time somewhere inside it.

The SPICE simulation is illustrated in Fig. 11.5. Note the log-log setting for the horizontal (frequency) and vertical (voltage, equal to impedance) axes.

So does the simulation line up with the calculations? Indeed so. The peaks at f_{psw} and f_{byp} are apparent, as is the resonance. This confirms that the calculations were done correctly. The simulation also shows that the peaks at f_{psw} and f_{byp} are dangerously close to X_{max}.

Selecting Bypass Capacitors

The engineering sidebar up in Chap. 3 discussed different types of capacitors, and the example above used two different types. The filter capacitor was 17.5 μF, which would probably get rounded up to 18 μF. (17.5 is not a standard value, but 18 is.) This capacitance is so large that a low-cost aluminum-electrolytic capacitor could be selected, unless physical space (or height) was at a premium. The bypass capacitors

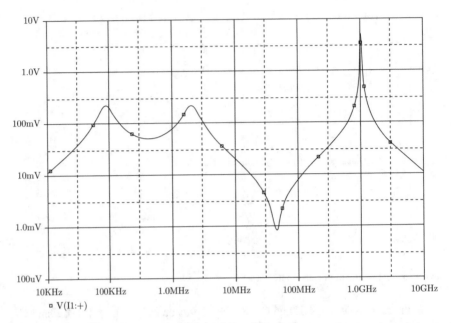

□ V(I1:+)

Fig. 11.5 Simulation of impedance of the power-supply network

were 0.01 μF, which is definitely small enough for surface-mount ceramic capacitors. The smaller the package size, the lower the ESL, so the smallest package size that is compatible with the surface-mount assembly process would be selected. This is often 0402, which is the smallest that can be reworked by hand, or could be something smaller.

One clear question is whether or not 125 capacitors make sense. It is not as ridiculous as it sounds, although it might be a little high. It is common for integrated circuits to have dozens of power and ground pins and therefore dozens of bypass capacitors around them. In "real life" the number 125 would be held up against the number of capacitors on the entire board. Chances are that it would line up well (there would probably already be nearly this many capacitors). If not, this would bear further inspection because of the very high risk of the resonance. More capacitors might need to be added, for example, or the selection of a slightly more expensive low-ESL ceramic capacitor. The example above highlights the value of the simulations and back-of-the-envelope calculations: The resonance calculation shows that this is a marginal design.

For an example of a design with actual decoupling capacitors, see Fig. 11.6.

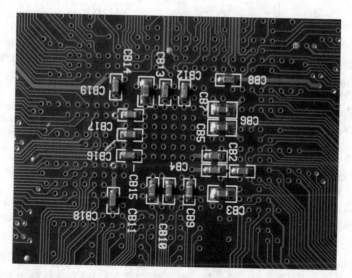

Fig. 11.6 Example of a "real" circuit board showing 18 decoupling capacitors. This is the bottom side of a board, below a medium-sized integrated circuit. The IC is surrounded by 18 capacitors. Stated differently, this one IC alone has 18 capacitors. If the board has 6 more ICs, the number of bypass capacitors is going to be close to 125

Appendix

1. Construct a power infrastructure for a simple design. The design has a target Xmax of 0.1 Ω, a rise time of 500 ps, and a power-and-ground-plane capacitance of 1500 pF.

   – Write down the last 4 digits of your Student ID number:

   Digit: $\overline{}$ A $\overline{}$ B $\overline{}$ C $\overline{}$ D

– Replace any instances of "0" with the last nonzero digit in your student ID number:

   Digit: $\overline{}$ A $\overline{}$ B $\overline{}$ C $\overline{}$ D

– Power-supply inductance is D*25 nH
– Inductance of bypass cap, L_{byp}, is C nH
– Resistance of bypass cap, R_{byp}, is $X_{max}/2 = 0.05 \ \Omega$

- Inductance of a single array capacitor, L_{cap}, is B nH
- Resistance of a single array capacitor, R_{cap}, is A/10 Ω

  (a) Compute f_{psw}, the point at which the power-supply wiring has too much impedance.

  (b) Compute C_{byp}, the amount of bypass capacitance needed.

  (c) Compute f_{byp}, the point at which the bypass capacitance has too much impedance.

  (d) Compute N, the number of array capacitors needed.

  (e) Compute C_{cap}, the capacitance value of each array capacitor.

  (f) Compute R_{array} and L_{array}, the resistance and inductance of the entire capacitor array, taking into account that there are N capacitors in parallel.

  (g) For the resonance between the capacitor array and the power and ground planes, compute the frequency of resonance and the impedance at the resonance.

2. Use the component values you calculated in part 1 to construct a SPICE simulation of the power infrastructure. For the array, you can assume the parallel combination of the R, L, and C. Run an AC sweep of the structure from 1 kHz to 10 GHz. One efficient way to do this is to create a 1 Amp AC current source and measure the voltage. Under these circumstances, the voltage is equal to the impedance. Do the results match your calculations in part 1?

Chapter 12
EMI/EMC: Design and Susceptibility

Background and Objectives

Up until now, the book has been about getting a design or product to work. More subtly, it has been about how to get a product to work *by itself*. This chapter explores what it takes to make sure that the design works with other devices and that other devices do not interfere with it. When this chapter is finished, you should be able to:

- Define electromagnetic interference (EMI) and electromagnetic compatibility (EMC)
- Understand sources of EMI, including the effects of circuit-board layout
- Understand the role of the chassis (or case) in reducing EMI
- Understand the role of cabling in EMI
- Understand the legal test requirements for EMI

EMI/EMC

When considering the effects of systems and devices on each other, there is terminology involved that is important to understand. When one device causes interference, it is called *electromagnetic interference* (EMI). The RF energy that it emits is called *radiated emissions* (if emitted into space) or called *conducted emissions* (if radiated onto a cable). The emissions can be anywhere from a few kHz to a few GHz, and, very generally speaking, conducted emissions are potentially worse because the device might be connected to power or network cabling that creates multi-hundred-foot antennas. The ability of devices to work together is called *electromagnetic compatibility* (EMC). These two concepts are strongly overlapping and so the subject is often called EMI/EMC. The likelihood of a device operating incorrectly in the presence of another device is referred to as its *susceptibility*.

© Springer Nature Switzerland AG 2022
S. H. Russ, *Signal Integrity*, https://doi.org/10.1007/978-3-030-86927-4_12

The long-and-short of it is that RF energy can be emitted by one device and can have measurable effects on another. Whether the RF energy is "unwanted" is like the definition of a weed – it depends on one's perspective. (To a wheat farmer, a rose bush is a weed.) This chapter is written from the perspective of unwanted RF energy, or at least unwanted effects on a neighbor.

There is a great deal that can be said on the subject of EMI/EMC. This chapter will highlight some important concepts, but you can find and consult other references if you need more information. Another important caveat is that this book is written from the point of view of consumer electronics. This book assumes you are not designing a cell phone for use, say, at a radar station. When electronics are designed to be operated near specialized high-power electronics, especially microwave and RF transmitters, very specialized techniques are needed that are outside the scope of this book. This is very important, for example, in the world of defense or aerospace electronics, where it is common for computer circuits to lie next to high-power pulsed-power circuits.

Also, the focus of this chapter is on EMI more so than EMC or susceptibility. It is about how to keep a product from *emitting* unwanted RF energy, from being an unintentional radiator. The good news is that all the steps that accomplish this, such as good chassis design, also reduce susceptibility. More good news is that if your product does not emit unwanted RF energy it is legal for sale...

The urban legends of EMI are entertaining. One of the more popular is the CD player that, when turned on, caused an airplane's navigational system to go haywire. This was apparently confirmed when the flight attendant had the CD player's owner turn it on and off repeatedly. The reason this legend is believable is that, for years, CD players were specifically listed in the back of in-flight magazines as prohibited from use. Another common example in the GSM cell phone era (wait, how many are too young for 3G?) was to hear buzzing when holding a phone near speakers.

What is causing these events to happen?

In the first story, the CD player is what is called an "unintentional radiator." It is not designed to transmit or emit RF energy, but it does. As we will see, the root cause of this is almost always the combination of bad circuit-board design and a bad chassis (either bad design, bad construction, or corrosion over time). At some point, the CD player had been tested and found not to radiate (assuming the CD player was legal for sale, more on that later), but for some reason, this CD player turned bad. At any rate, there was some source of RF energy at the frequency used by the navigation system, and the RF energy was able to couple out of the chassis of the CD player. (Chassis is a fancy word for case or package.) This is the recipe for radiated emissions – a source of RF energy and a coupling path to the outside world. For an unintentional radiator, either the source of energy or the coupling path (or both) are unwanted. One very likely scenario is that the unwanted energy was sent out over the headphone cords (this was pre-Bluetooth, after all) and the headphone cords formed a good antenna. RF energy is far too high in frequency to be heard by the human ear.

The story of the cell phone is murkier. First, the cell phone is an "intentional radiator" meaning that it has a radio transmitter in it. A modern smartphone has at least three, one for the cell tower, one for Wi-Fi, and one for Bluetooth. These are covered under slightly different versions of the regulations, and must be tested to make sure they only radiate at the frequencies they are intended for. Intentional radiators create problems when, as is the case with 3G and 4G LTE, their intentionally used frequencies overlap with, for example, the frequencies used by cable television. So one problem is that the cable company cannot tell the cell phone operators to stop using the frequencies.

Second, cell phones put out a lot of power, relatively speaking, and so can generate a substantial E field, especially when held near other electronics. Again, this high-power level is due to the fact that it is emitting RF energy on purpose, and (in the case of a cell phone) the signal has to travel several miles. The underlying physics is the same – source of RF energy plus signal path – but in this case, it has been optimized for maximum efficiency.

Third, the chassis of an intentionally radiating product has to have a hole in it for the intentional radiation to get out; this opens up the prospect of unwanted noise getting out with it.

So with all this in mind, how are products designed to avoid unwanted radiated and conducted emissions, and what are the legal requirements?

Circuit-Board Design

The first step in preventing EMI is to avoid having a source of unwanted RF energy. This is called *source suppression* and, as one might expect, is the most effective way to deal with the problem. The good news is that all of the layout and design techniques discussed in this book have this effect – good layout and good signal integrity tend to eliminate unwanted RF energy. Think about it – no reflections, no crosstalk, no unexpected resonances in the power and ground plane. A good design will simply have fewer radiated emissions.

There are some aspects of radiated emissions that can still crop up, however.

For example, under what circumstances does a high-speed signal tend to leave the circuit board? One way that this can happen is if the signal is near the edge of the circuit board, as illustrated in Fig. 12.1.

Consider signal 1 and signal 2 in Fig. 12.1. The E field of signal 1 is well confined to the circuit board, but the E field of signal 2 fringes out past the end of the board. When this happens, the E field formed between the signal and ground fringes out into free space. There might be a chassis there in free space, in which case the chassis might pick up some of the signals and reradiate it. (By the way, the word "fringing" is "fringe + ing" and not "f-ringing". It comes from the word "fringe" which is, you know, the little tassels on the ends of fabric.)

This leads to the *10H rule*: For a height H above the ground plane, keep signals at least 10H away from the sides of the board. The 10H rule is illustrated in Fig. 12.2.

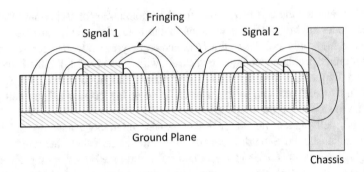

Fig. 12.1 Fringing of a signal near the end of a circuit board

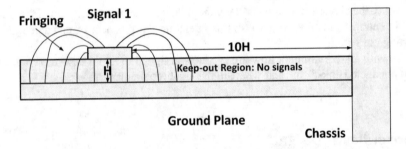

Fig. 12.2 10H rule. The rule is implemented by not routing any signals within "10H" of the board edge. This is drawn to scale – the horizontal spacing is 10 times the vertical spacing

As is the case with crosstalk, some designs even add a grounded guard rail around the edges of the board, although this can be difficult to add and requires nudging every time the board outline changes.

Another problem area is the power and ground planes. These can create EMI issues for two reasons. First, switching transients tend to get dumped onto the power and ground because of the need for real power to create the transient. A common example is ground bounce. Second, they form very large structures and so tend to form very efficient antennas. A large transient on the ground plane tends to become a large EMI problem. One common solution is to invoke the "10H" rule on the power plane. If the power plane is "10H" away from the edge of the ground plane, the fringing between the power and ground plane stays inside the circuit board.

Finally, any place that the board contacts the chassis or cabling is a point that unwanted energy can efficiently exit the system and enter the outside world (where it can make a design no longer legal for sale and make the company's future questionable). Think of the chassis and connectors of the unit as a type of quarantine – they have to hold in the unwanted RF energy that remains after creating a good design. In turn, this breaks down into two types of signals, ground and everything else. Ground exists in the form of the chassis and the shields on connectors (and most cables). "Everything else" is the signals that run on the connectors. But first, we need to learn more about chassis and grounding.

Engineering Notebook – Mounting Holes, Cable Shields, and Grounding

There are always two aspects to the design of a chassis and of cables and connectors, electrical and mechanical. The two aspects unavoidably overlap (and so the rivalry between electrical and mechanical engineers is inevitable).

Since electrical engineers are reading this book, one can start with the mechanical. There has to be a way to hold the circuit board inside the chassis, and so the chassis is designed with a combination of pegs and screw-holes to match the circuit board. The number and spacing of the mounting features depend on the requirements of the product (e.g., does it have to withstand a drop onto a concrete floor?) and on the design of the chassis. Some features prevent the side-to-side motion of the board, such as a peg or metal flange that sticks up through a notch in the board. Other features also prevent up-and-down motion such as a screw. The screw typically attaches to some feature on the chassis, such as a raised hole, in which case the board is actually contacted on both sides. The combination of holes and pegs is a mechanical issue – the board has to be held down in a way that meets the product's mechanical requirements.

A typical screw hole is illustrated in Fig. 12.3.

As one can see in Fig. 12.3, screw holes require a special circuit-board design because of the obvious need not to put components where the screw and the corresponding opposite mounting feature will touch the board. In the figure, the mechanical engineers created a plain hole in the chassis (more accurately, they

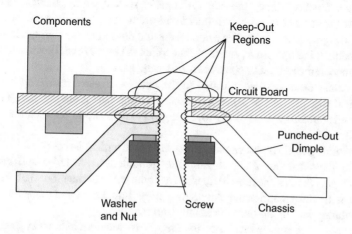

Fig. 12.3 Typical screw hole

designed a chassis such that the chassis manufacturer punches the dimple and then drills a hole) and then uses a nut to hold on the screw. The washer is there to make sure the screw makes good electrical contact. The alternative to using a washer and nut is to use a threaded hole in the chassis. (Using threaded holes has advantages and disadvantages that mechanical engineers can explain.)

Because screw-hole placement creates circuit-board layout constraints, the mechanical engineers typically create the board outline and the screw-hole placement and then provide the keep-out regions and circuit-board outline to the electrical engineers to start circuit-board layout. The problem is that screw-hole placement has electrical ramifications.

Note that the screw might be insulated from the rest of the board or it might be connected to ground. If the top of the board (under the screw head) or the bottom of the board (next to the dimple) is grounded, then attaching the screw will cause the screw to be grounded. (Extra solder may be added, or a star washer can be put under the screw to improve the electrical contact.) Likewise, if the sides of the hole are grounded (if the hole is a plated hole, also known as a plated via) the screw might be electrically connected to ground. If the top of the board under the screw, bottom of the board next to the dimple, and sides of the hole are left bare, then the screw will be insulated.

Another design aspect that the mechanical engineers typically "own" is connector placement, because the connectors have to line up with the holes in the chassis. The connector that goes onto the circuit board also has its own mechanical issues. Specifically, the connector has to withstand the mechanical forces of insertion and removal in such a way that the wires (and solder) carrying the signal do not break.

A typical way of creating a mechanically sturdy connector is to construct the ground shield of the connector with through-hole pins. That is, most connectors are through-hole parts, not surface-mount. (Very high-speed connectors, such as SATA and HDMI, need surface-mount connections for the signals, and so the grounded shield is through-hole.) The important point is that the pins serve an important mechanical function – they provide *strain relief* so that the mechanical forces are transmitted to the fiberglass circuit board and not to the signal wires or, worse, the solder. Conversely, if you choose a connector that does not have through holes, you will probably wind up with a connector that comes off every time someone unplugs a cable. I have removed many connectors this way, alas.

Some connectors are also attached to the chassis. Besides providing additional mechanical strength, this is sometimes done to dump heat more efficiently into the chassis and sometimes done to improve the ground connection between the connector and the chassis.

So, returning to EMI and cabling, what are the electrical issues?

By far the most important is this: The grounded screw holes and grounded connector pins (and connector-shield pins) are connected to the ground of the circuit board, usually the ground plane. Any return current on the ground plane near one of these features can hop up onto the feature and radiate.

So how is this prevented? The first step is to route signals away from these features, away from grounded pins or holes. In fact, the situation is essentially

Fig. 12.4 Example of moating

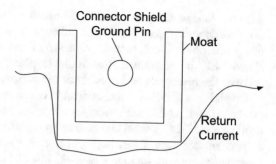

identical to the problem of signals near circuit-board edges, and so this implies at least a 10H spacing. To be clear, this implies that signals and power planes must be at least 10H away from a screw hole or a connector ground-shield pin.

The second step is to consider *moating*. (For nonnative speakers of English, a *moat* is the lake that is built around castles to keep out unwanted invaders.) This is actually the place where one really does put slots in ground planes. Keep in mind the first step, though – there should not be any signals near these slots. By selectively adding slots in the ground plane, one can cause return currents to divert around a mounting feature while maintaining a solid electrical connection between the ground plane and the feature. This is illustrated in Fig. 12.4.

The picture may be a little confusing. Figure 12.4 is a picture of a ground plane. Everything is a solid sheet of copper, except that there is a hole and a U-shaped "moat" has been cut where copper has been removed. In Fig. 12.4, the moat diverts stray return current around the ground pin of a connector shield. The moat creates a high-impedance barrier and so the current steers around it. The return current has no reason to go onto the connector pin since the current is the return for some other signal. In other words, the return current has no need to cross the "drawbridge" that connects the pin to the rest of the board – the return current is returning a signal that runs between two points on the board. On the other hand, the connector pin remains well-grounded; it still has very good electrical contact to the ground plane. This strategy will work both for mounting-screw holes and for connector shield ground pins.

Chassis

The chassis is the last barrier of containment for any radiated signals, and so its design is critical for success. Most chassis are metal, although as devices continue to shrink, plastic is coming more into vogue. If a product is being designed with a plastic case, there are still options, such as lining the inside of the plastic case with a metallic coating.

The chassis acts like a Faraday cage, a solid metal structure that completely electrically isolates the interior from the exterior. In fact, if the chassis were solid there would essentially be no radiated emissions. But there are two reasons why this can never be. First, there have to be holes, both for ventilation and to let interesting things like power cords, network cables, and user interfaces come out. Second, the chassis may be assembled from multiple pieces, in which case the connection between the pieces can be imperfect or can corrode.

Obtaining a good electrical contact between two non-corroded pieces of metal can be difficult, and in fact, becomes a mechanical issue (again). There has to be a positive force holding them together, and it is preferable that the contact is made over a wide surface area. Likewise, corrosion is more subtle than one might think. Corroded aluminum, for example, looks just like non-corroded aluminum. (In fact, when you see a piece of aluminum it is almost always corroded.)

So what about the inevitable holes? The rule of thumb is that the longest wavelength that can radiate out of a hole is equal to the longest distance across the hole. So a slot (long, skinny hole) will allow RF energy of lower frequency out than a round hole. This is why ventilation is usually provided through a collection of round holes – round holes minimize the maximum-distance constraint for a given amount of surface area. In other words, round holes let through the most air and yet block the most RF energy.

What is the maximum frequency of the signal energy inside the chassis? You should know this by now – it is the knee frequency. (Another example of why knee frequency is so important!) So the holes or slots in the chassis must have maximum length well below the wavelength of the knee frequency. Fortunately (for now) this constraint is usually met with little difficulty. 1 GHz has a wavelength of about 1/3 of a meter, or about 1 foot.

There is another possible issue with screws and their holes. Two screws can create a loop (formed by the ground plane, the two screws, and the chassis). This can create a loop antenna that can radiate efficiently. Keeping return current away from the screw holes will help, which is another example of how moating can be helpful. Also, it will help to make only one screw hole grounded – there will not be a second screw to conduct an eddy current.

It is important to review the problem with fringing – signals near the chassis can fringe out onto the chassis, and signals near a screw can fringe out into the screw. This highlights the need to route signals away from the chassis, which normally boils down to keeping signals away from the edge of the board. In fact, only signals that actually enter or exit the board (i.e., signals going out to connectors) should be near the board edge. This is normally the case, as common-sense signal routing will tend to route internal signals near the internal parts of the board, and only signals traveling to and from connectors are routed to the board edge. Importantly, avoid the temptation to use the outer part of the board as a highway for signals that need to spread out due to routing congestion.

One veteran EMI expert that I worked with saw instances where the signal energy coupled directly from the board up into the chassis and connector grounds. Again, the 10H rule will minimize the likelihood of this.

Cabling

Cables create headaches in three basic ways.

First, as noted above, the cable's grounded shield can create a very large antenna if the ground connection has stray return currents. Again, careful signal routing, moating, and the 10H rule will reduce this issue.

Second, return currents on the cable can create problems. This is subtle, but can create large headaches. Consider a single-ended signal (a signal that uses ground for its return current) routed over a cable. The return current for the signal can either travel back on the grounded signal on the cable (i.e., it can use the ground wire on the connector for its return), or it can follow the path through the chassis ground of the receiving device, through the power wiring, to the chassis ground of the transmitting device. The second path has considerably more inductance, but that also means that it is traveling over a gigantic loop and therefore forms a very efficient antenna. This structure is sometimes called a *ground loop*. This is illustrated in Fig. 12.5.

The arithmetic is surprising. The efficiency of the antenna increases linearly with the loop area, and the inductance of the antenna increases with the natural logarithm of the area. Thus, the radiated emissions go up as $A/\ln(A)$ where A is the area of the loop. So the bigger the loop formed by the power cords, the more it will radiate.

The solution is to add inductance creatively. By putting a ferrite bead (i.e., a cylinder of ferromagnetic material) around the cable, the return current traveling through the power wiring (which lies outside the ferrite bead) "sees" the inductance of the bead. Conversely, the inductance of the signal and the return current traveling inside the cable tends to cancel. So the addition of a ferrite bead forces almost all of the return current to travel over the cable instead of the power wiring.

Companies that manufacture ferrite beads are very excited about this. It is always good business practice to manufacture products for which the demand is driven by the laws of physics. These companies will cheerfully give you samples of all shapes and sizes of ferrite beads. However, if you pass radiated-emissions testing with a ferrite bead on a cable, you will have to ship the product with that specific ferrite

Fig. 12.5 Example of a ground loop

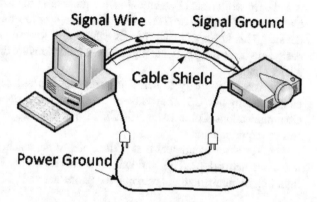

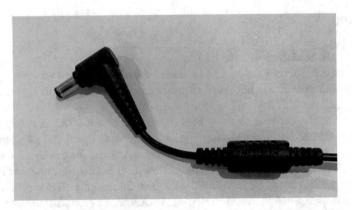

Fig. 12.6 Example of the ferrite bead. The "bead" is the cylinder-shaped object on the right. They are often seen on laptop power cables and projector cables

bead on that specific cable. In other words, every time you ship a product, you have to include a specific cable with a specific ferrite bead on it.

Sometimes this is essentially unavoidable. This is why almost every projector ships with a VGA cable with a ferrite bead on it. (The bead is the cylindrical lump around the cable near one of the connectors.) The rule of the thumb, though, is to first try to pass radiated-emissions testing without the ferrite bead. An example of a ferrite bead is shown below in Fig. 12.6.

This leads to the next topic, namely radiated-emissions testing.

Legal Certification: Standards and Test Requirements

Have you ever thought about what it would take for a product that you designed *not* to be legal for sale? It turns out there are two basic ways to do this, at least in the US.

The first, and most obvious, is to make a product that is unsafe. Fortunately, there is a clearly written set of electrical-safety guidelines that we can all follow. (This is the source of the UL or ETL mark you have seen on practically every electrical device.) This keeps our own homes safe and, if you think about it, lets you as a design engineer sleep better at night. These guidelines are discussed in more detail in a later chapter.

The second is to try to sell a product that has not been certified for radiated emissions. In the US, the body of requirements is maintained by the Federal Communications Commission (FCC). In our line of work, there are three basic sets of requirements.

One set is for equipment that is designed to be attached to the public switched telephone network (PSTN), and that set is called Part 68. Part 68 mainly deals with lightning immunity and not crashing the phone network. This lies mostly outside the scope of this book, but engineers should know about it in case they go to work for a company that designs phone-network equipment.

One set is for unintentional radiators, part 15 subpart B, and another is for intentional radiators, part 15 subpart C. This is the set most relevant here. There is also a distinction for Class A, commercial equipment, and Class B, home equipment; Class B has tighter limits.

To make a long story short, the requirements call out the need for a laboratory to test a product and certify that it meets the requirements. The laboratory is normally a test lab near the designing company, and test labs that perform this testing are fairly common. Most companies have a "go-to" lab that they use frequently. (In medium-sized companies and larger, there is often a full-time staff member that coordinates the testing with the lab.) This testing is usually called "FCC testing."

The European standard is more all-encompassing. In addition to radiated emissions, products must be tested for susceptibility, electrostatic discharge (ESD) immunity, and "dips and interrupts" (interruptions to the power-supply input). This testing, in addition to safety testing, is required to get a "CE mark."

To conduct radiated emissions testing, the lab uses calibrated antennas connected to a spectrum analyzer. (A spectrum analyzer produces a display of signal energy versus frequency.) The antennas are placed a very specific distance away from the product (usually 10 meters, but the lab may use a different distance and correct for it). The product is placed on a turntable, all cables that can attach to the product are attached, and the product is powered up. The product should then attempt to use all of the cables and put outputs on each of them.

The technician making the measurements will rotate the product and adjust the antenna's polarization to get a maximum reading. A typical result is a flat measurement with several spikes at specific frequencies. If any of the spikes exceed the legal limit, the product fails.

If one has a failing result, one can often look at the frequencies of the spikes and figure out which signal is to blame. For example, if is a multiple of the frequency of HDMI (around 740 MHz), then HDMI is likely to blame. If it is a multiple of the clock frequency of the main processor, the design may be in trouble because that is very hard to reduce (the clock drives the processor which drives the entire board).

If a product passes, the lab can usually certify it both for US and for European purposes. Obtaining CE-mark certification, fortunately, does not require testing in Europe. Once the lab certifies that the product meets the legal limits, then one hurdle for selling the product has been cleared.

Needless to say, I have a good FCC testing story, which makes a good place to wrap up the chapter...

Engineering Notebook – Get Down Here Right Away

We had finished a redesign of a circuit board and had changed the main power supply from a linear regulator, which has very little RF energy, to a switch-mode power supply, which pumps out a lot of RF energy and dumps it onto the ground plane. We tested the board and it failed FCC testing by a huge margin.

When I did a redesign, I knew the main culprit was the power supply, so I elected to employ moating. The closest thing to a ground on the board was a screw hole, so I added a moat around the screw hole that extended to the power supply. My hope was that the noise from the power supply would be dumped onto the screw, in effect.

When we got the board back, I tested it with a near-field probe, which is a scientific way of waving a coil of wire around to see what noise is on it. The near-field results were significantly worse, and I thought that I might have to trade in my "signal integrity expert" reputation for a bag over my head.

We took the design to FCC testing to get a scientific picture of how bad it was. The technician started testing the unit, and it started passing! So I called back to the office and told our staff FCC-certification person to get to the lab right away – we were passing! The product received FCC certification that day.

The moating strategy had actually worked. The bad results from the near-field probe were due to the fact that the noise was confined to a small area of the board, instead of radiating out of the entire ground plane. The noise was more concentrated but the point is that it was confined. The FCC is perfectly OK with this, of course, because the product was not radiating energy out to interfere with other products.

Appendix

1. Define the following terms:

    (a) EMI
    (b) Radiated emissions
    (c) Conducted emissions
    (d) EMC
    (e) Ground loop
    (f) FCC
    (g) PSTN
    (h) FCC Part 15
    (i) FCC Part 68
    (j) CE mark

2. Contrast intentional and unintentional radiated emissions. Give examples.
3. What circuit-board layout rule is used to keep RF emissions away from mounting screws?
4. What two functions do pins or mounting screws serve when used as part of connectors?
5. What circuit-board layout rule can be used in the vicinity of connector-shield pins?
6. What structure is often added to cables to prevent unwanted radiated emissions from ground loops?

Chapter 13
Electrostatic Discharge

Background and Objectives

We have all touched a metal object after walking across a carpet and received an electric shock. This shock is officially known as electrostatic discharge (ESD), and what is a discomfort for us can severely damage electronics. This chapter studies the origin of ESD and strategies to make designs resistant to it. When this chapter is finished, you should be able to:

- Explain how ESD events occur and how they are measured
- Explain what effects ESD can have on electronics
- List some design techniques and components that can manage ESD

What Is Electrostatic Discharge?

Electrostatic discharge (ESD) occurs when some charge that has been accumulating discharges suddenly. It is as if a capacitor is suddenly shorted out. An electric charge can accumulate on any insulator. When the charge-carrying insulator is brought near an electrical conductor, the charge can capacitively couple (or conduct) from the insulator to the conductor. Again, a common example is touching a doorknob or car handle and getting a jolt. In this case, you are the insulator (the charge literally accumulates on and in your body) and the breakdown occurs during the moment of your touch.

The problem with ESD is that electronic components, especially miniature modern components, are very susceptible to ESD-related damage. If the jolt from an ESD event is enough for you to feel, it is far more than the amount needed to destroy most transistors. The result is that electronic products, especially products designed to be handled or touched frequently by humans, must be designed for ESD

© Springer Nature Switzerland AG 2022
S. H. Russ, *Signal Integrity*, https://doi.org/10.1007/978-3-030-86927-4_13

immunity. This design occurs at multiple levels, from individual chips all the way up to an entire product.

ESD is typically measured in volts, but this is quite misleading. At the end of the day, ESD is a charge event, the rapid dissipation of a large amount of electric charge. Measuring the voltage is a way of estimating the amount of charge that was collected before the ESD event occurred. As easy as it is to make the measurement, it is like measuring how high the dam was before it burst – the waterlogged people living downhill from the dam don't really care.

A human being can accumulate enough electric charge to be several kilovolts above ground. (From an ESD modeling standpoint, a human being is a 100 pF capacitor in series with a 1500 Ω resistor, so several kilovolts is still only a few nanofarads of charge.) When a human touches a doorknob and gets a jolt, the resulting waveform typically has a rise time of 200 ps to 1 ns and a peak current of 30 Amps. Because of the rise time, the event has RF energy out to several GHz.

So a typical ESD event combines an RF-energy component, which is therefore like an EMI/EMC event (Chap. 12), with a current surge of a few amps to a few dozen amps.

It is very important to note a special class of ESD event, lightning. Yes, lightning is a form of ESD, albeit on a relatively colossal scale. The voltages and currents involved in lightning are much greater, of course, and lightning is so energetic that it literally emits X radiation. But the techniques used to manage ESD will also work with lightning, fortunately.

ESD events can cause damage through one of four different ways. First, the current surge flowing directly through a component or input/output can directly damage it. Fortunately, even a small series resistance or parallel capacitance can substantially reduce the damage. Second, the current surge can cause a ground reference to develop a very high voltage (or a power rail can cause a very low voltage). Inverting power and ground can cause major issues. For example, CMOS integrated circuits can enter a state called latch-up which can permanently destroy them. Latch-up occurs when the large doped areas of internal circuits, such as the N-wells and P-wells that form logic gates, form an intrinsic silicon-controlled rectifier that short circuits. In other words, the chip effectively becomes a large short circuit. Third, the current surge can cause EMI (which is essentially the same as crosstalk in this context), which can cause signals to be disrupted. This is usually not destructive but could cause operational errors. Fourth, the static E field, due to a very high voltage before discharge, can cause damage. It turns out this fourth case is very rare – normally something on the circuit board discharges first.

Where Can ESD Occur?

ESD mainly occurs in three places.

First, it can occur in any place where humans touch. This often includes structures like control knobs, buttons, and touch screens. So any sort of human-interface control usually needs some sort of ESD protection.

Second, it can occur in any place where humans plug-in cables. What actually happens is that the human charges up the cable, which then discharges when plugged in. This is an important reason for having ground shields on connectors – the ground shield makes contact first and can usually dissipate the charge. In some standards, such as SATA, the ground pins on the connector are made longer than other pins so that they can also dissipate the charge before sensitive signal wires make contact.

Third, it can be carried into the device by cables that have already been plugged in. In other words, any signal carried onto (or off of) the board by a cable can potentially carry an ESD event. If the cable is a long one, such as a power, network, cable-TV, or telephone cable, it can also potentially carry a lightning event.

This is important to understand, as it drives the analysis of which signals require ESD protection. It also drives the development of testing standards in various industries.

ESD and Lightning Standards and Testing

In terms of ESD, whether there are official standards depends on where your product is sold. In Europe, ESD testing is required in order to get a CE mark. (Keep in mind that devices to be sold in Europe will need a CE mark.) In the US, it is more haphazard. Each company develops its own ESD standards. At the risk of being preachy, it is probably a good idea for a company to simply adopt the European standard since it is widely used and seems to work. ESD standards are clearly important for consumer devices, as they are routinely touched and handled by statically charged humans.

ESD testing is typically carried out by placing the device under test on an electrically isolated table and then probing it with a high-voltage probe. The probe is designed to charge to a calibrated high voltage (e.g., 15 kV) and then discharges if it comes near a place on the device that is susceptible.

Consider the front panel of a piece of stereo equipment. As the wand passes over most of the front panel, nothing happens; there is too much plastic in the way. As the wand nears a knob or button, it might discharge through the metal of the button or through the gap in the plastic that lets the shaft of the knob stick out.

The behavior of the device, such as whether it is permanently damaged or hiccups and then recovers, is then recorded as the wand is passed over all parts of the device. This determines whether the device passes or fails testing.

The probe is actually dangerous (one company calls their model "Minizap") and so it is important to read and follow directions and standards closely. For example, no one should perform ESD testing alone. An extra observer in the room is needed to see if a medical emergency occurs. Many of the labs that perform radiated-emissions testing can also perform ESD testing (since both are required for a CE mark) and so this is often an attractive option as opposed to maintaining your own dangerous test equipment.

Lightning immunity requirements are dependent on the interfaces that the product uses. Phone lines have their own standards, FCC Part 68. Ethernet requires a transformer and a 1500-Volt capacitor. AC power is required to have DC isolation according to UL standards. It is important to identify which standards are relevant and then design them accordingly.

Lightning testing is only slightly as scary as it sounds. Fortunately, there are commercially available lightning-test systems, and many of them are designed for relevant standards, such as FCC Part 68.

So with all of the potential for mayhem, how can devices and products be made ESD-resistant? The answer boils down to two basic approaches, extra components and design techniques.

Components to Manage ESD

As we have seen, ESD causes a large, fast surge of current to rush into a circuit, usually with an accompanying sharp rise in voltage. ESD circuits are designed to carry the current away from sensitive devices (normally to ground) and to dissipate the energy of the surge of current.

One important lightning-protection device is a simple fuse. On the time scale of a lightning event, fuses are glacially slow, but they can disconnect a device from a line carrying a transient, and so have a nearly infinite power-dissipation capability. Keep in mind that there are self-resetting fuses, such as Polyfuses or other polymer-based devices, which are a good choice if a piece of equipment needs to operate remotely. On the minus side, Polyfuses are even slower than conventional fuses – they rely on a material heating up and expanding.

One very common ESD-management device is a simple diode. A typical arrangement is to place one diode between ground and a signal, and another diode between the signal and the power rail. This structure is illustrated in Fig. 13.1.

If the signal falls below ground, for example, the first diode begins to conduct and will clamp the signal to the forward voltage of the diode (0.7 V if it is a silicon diode). So the signal cannot fall below -0.7V; it is clamped in place by the diode. Likewise, the signal also cannot rise above Vcc + 0.7V.

Fig. 13.1 Typical use of diodes in ESD management

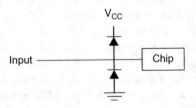

This is an extremely common ESD circuit. It is used in USB, for example, and can be easily constructed inside integrated circuits. This diode-pair arrangement is so common that there are often single-package components containing both diodes, like the BAV-99 diode pair. The circuit has only two potential drawbacks. First, it only works if the diode can dissipate the energy of the ESD event without being destroyed. Second, it may add capacitance to the circuit (the capacitance of the diode in its non-conducting state).

Two more classes of ESD devices are often used if greater amounts of energy dissipation are needed.

The first one is designed for *clamping* behavior, very similar to that of the diode structure discussed above. These devices are open circuits until the input voltage exceeds the breakdown voltage. Once the breakdown voltage is reached, the device clamps the signal to the breakdown voltage. Typical structures that perform this function are called transient voltage suppression (TVS) diodes. They can dissipate a substantial amount of energy and can turn on extremely fast, on the order of picoseconds if the package inductance is low. However, they only clamp the input to the breakdown voltage; a substantially high amount of voltage might still get through. The keys are to select a device with a suitably low breakdown voltage and, of course, to study manufacturers' datasheets and application notes.

The second one are *crowbar circuits*. These devices turn into short circuits in the presence of an overvoltage condition. Notice the difference – transient voltage suppression devices remain at the breakdown voltage when turned on, and crowbar circuits turn into short circuits. There turn out to be three basic types of crowbar devices.

The first crowbar circuits are semiconductors, such as DIACs, SIDACs, and SIDACtors, which operate a lot like silicon-controlled rectifiers or TRIACs. These turn into short circuits in the presence of large positive and negative voltages. They are often used in FCC Part 68 (telephony) applications because they can absorb lightning transients (both positive and negative) in a single component.

Crowbar devices can be used to protect a circuit board from a runaway power supply if that is a likely failure mode. For example, in one of my designs, the voltage feedback element in the power supply was unreliable and the power supply would go into a full open-loop mode that maxed out the output voltage. When that happened, the crowbar device would turn into a short circuit and force the power supply into shutdown.

The second crowbar circuits are metal-oxide varistors (MOVs). These are often found in power strips and other similar types of power-protection devices. They can handle substantial transients and dissipate quite a bit of energy, but they have substantial disadvantages. Specifically, as they dissipate transients, they can become more conductive in their off state. Eventually, they reach a point where they do not turn completely off; they leak current which means they heat up. They can even reach the point where they burst into flame. You read that correctly – the safety device in your power strip can, under certain circumstances, burst into flame. They are actually less likely to develop this failure mode if they are hit with fast, large transients, and so they ironically actually work better if they are not used in combination with other energy-dissipating structures.

The third crowbar circuit may sound surprising but actually works; it is a spark gap. When a spark forms, it is actually a highly conductive plasma, a soup of ions. If a spark forms in the open air, we can see the characteristic bluish-white light, and it also can lay down or coat the area where it strikes with carbon. This coating occurs because it pulls carbon from the carbon dioxide in the air. However, if the spark forms inside a sealed container filled with inert gas, the spark can be struck repeatedly with no degradation. This structure can dissipate quite a bit of energy.

One common ESD or lightning-suppression structure is a parallel (shunt) capacitor. This structure exploits the simple fact that the capacitor is an AC short circuit. It is used in Ethernet, in concert with a transformer (discussed below), to provide lightning immunity.

Another common lightning-suppression structure is a parallel (shunt) inductor. This may seem counter-intuitive (since a capacitor will also work) but the inductor is an open circuit at radio frequencies and a short circuit at DC. A parallel inductor will let radio signals pass but short a large DC lightning impulse to ground. This is used at the input of cable set-top boxes, for example.

Series resistors or inductors can also serve a useful purpose, namely slowing down the current surge until other devices, such as TVS devices or capacitors, can shunt the energy away.

A useful technique for lightning immunity is to add circuits that provide DC isolation. The most common example is a transformer, which only carries AC voltages across it. The DC current pulse does not pass through the transformer. Most main power supplies (power supplies connected directly to AC power) use transformers for this purpose, as does Ethernet. In fact, in both cases (AC power and Ethernet), the transformer is required by safety standards. Even if the power supply is destroyed by the lightning surge, the rest of the system is not and can therefore be repaired. It also makes the system much less likely to catch fire, which is a laudable goal even if the system is destroyed.

Another DC-isolating circuit is an opto-isolator, which is an LED (or infrared emitting diode) in series with a photodiode. The signal is conducted optically through the opto-isolator instead of electrically. They are commonly used as feedback elements in power supplies where DC isolation is needed.

In terms of lightning, there are two basic ways it can cause harm. One is for the actual lightning strike to be conducted through the wiring. This is really bad. Fortunately, the second is much more common – normally a "lightning strike" inside a house is either only fraction of the actual strike (with the rest of the current of the strike finding a lower-impedance path to ground) or is caused by the electromagnetic pulse (EMP) of a lightning strike. "EMP" is being used loosely here – it represents the fact that many times the lightning damage inside a home is not caused by the lightning bolt itself but from the burst of EMI that is caused by the bolt.

Design Techniques to Manage ESD

There are two schools of thought for managing ESD, and fortunately, they overlap.

The first school is to design the plastic of the chassis to minimize the likelihood of having an ESD event. One can think of it as designing to pass the ESD tester system discussed earlier, but in fact, one is actually hardening the device to prevent ESD events up to 10–15 kV. This requires making the plastic thicker in areas where an arc is likely to strike, such as the corners of display screens and near where buttons and knobs penetrate the front panel.

The second, more conventional, school of thought is to design the board to shunt the energy away from sensitive components. The key is to design in layers. The outermost layer, the ones closest to the actual connector or point of entry of ESD, is typically a combination of spark gaps, high-voltage capacitors, and fast, high-energy-absorbing semiconductors (in that order). Spark gaps are the most rugged and therefore, if they are used, come first. The semiconductors do not have the energy-dissipating ability of spark gaps, but turn on extremely fast. If a circuit only needs minimal ESD protection, a pair of silicon diodes can be used. The next layer is a series element, such as an inductor or resistor. The goal of this layer is to slow down the transient to the point where the outer layer can finish cleaning it up. An optional third layer is to add bypass capacitance to filter out any remaining noise, but this may add too much capacitance to the signal.

In terms of circuit-board layout, most of the design practices we have already studied will help, such as keeping lines short and keeping the ground plane close to minimize inductance. Careful moating may also be helpful. (See Chap. 12 for a detailed discussion on moating.) The goal of the moating is to route ESD current that is dumped onto the ground plane (e.g., by a crowbar device) away from sensitive devices and towards either chassis ground or the power supply. If a design has a separate chassis ground, the energy-absorbing elements should be tied to chassis ground and not logic ground.

One common strategy is to add a grounded guard band around the edge of a board on the top and bottom sides. It is unclear, though, whether this might make ESD testing worse as it increases the likelihood of arcing under testing.

Appendix

1. Define the following terms

   (a) ESD
   (b) Clamping circuits
   (c) Crowbar circuits
   (d) TVS
   (e) MOV
   (f) Opto-isolator

2. Identify some ESD practices used for human inputs such as pushbuttons
3. What type of ESD suppression device is commonly used on circuit inputs, such as USB connections?
4. Which common technical standard requires both a transformer and a high-voltage capacitor to provide lightning immunity?
5. Name an example of a device that becomes a short-circuit in the presence of high voltage.
6. Name an example of a device that becomes an open circuit in the presence of a high current.
7. Why is ESD testing dangerous?
8. Is ESD testing required for product approval in the US? Is it required for product approval in Europe?

Chapter 14
Clocks, Jitter, and Eye Diagrams

Background and Objectives

We conclude our tour of signal integrity by focusing on clocks and clock sources. Clock sources turn out to be the demanding, flamboyant, dramatic center of attention of electronic design, requiring extremely careful layout and often causing significant design headaches. When this chapter is finished, you should be able to:

- Define clock jitter and clock skew
- Understand how to measure clock jitter
- Understand clock sources such as voltage-controlled oscillators and phase-locked loops
- Understand the origins of clock jitter and how to minimize it
- Understand eye diagrams and their role in signal integrity

Clock Jitter and Clock Skew

The first term that needs to be understood is the *clock*. The average person thinks of a clock as a face with two hands on it that ticks. This type of clock actually has two components, a source of periodic events and a mechanical system to count the events. In microprocessors, there is often a similar structure that is called a real-time clock or timer.

In this chapter, when we speak of a "clock," we are talking exclusively about the source of periodic events, usually a square wave that oscillates at a very specific frequency.

We have seen a lot of indications of bad signal integrity, such as crosstalk, ground bounce, and reflections. Each of these can be measured and used to grade the quality of a layout or design. But what about clocks? What figures of merit describe the quality of a clock signal? It turns out there are two, skew and jitter.

© Springer Nature Switzerland AG 2022
S. H. Russ, *Signal Integrity*, https://doi.org/10.1007/978-3-030-86927-4_14

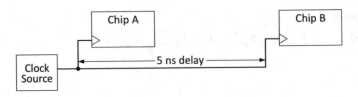

Fig. 14.1 Example of clock skew

Consider a clock that has to go to multiple digital chips, such as a memory clock going to several memory chips. The physical layout of the circuit board is such that the clock will arrive first at chips that are a shorter distance away from the clock source. *Clock skew* is the time difference between the earliest-arriving clock and the latest. It is illustrated in Fig. 14.1.

In this example, the clock arrives 5ns later at Chip B than Chip A. If Chip A is sending data directly to Chip B, this can cause serious problems. For example, data coming out at the start of a new clock cycle at Chip A can arrive at Chip B before the clock from the previous clock cycle. If Chip B is sending data to Chip A, the problem may be worse – Chip A will clock the data in 5 ns earlier than Chip B sends it.

Clock skew can cause incorrect clocking. For example, setup or hold times of flip-flops might be violated if clocks arrive relatively early (earlier than the accompanying data) or relatively late. Given enough skew, flip-flops may also clock in incorrect data.

The clock period of a digital system is constrained by a series of steps. The clock edge arrives at one flip-flop and new data is clocked into it. This data appears at the output of the flip-flop and propagates through a network of combinational logic. Some time after the network of combinational logic has settled to its final state, the next clock edge arrives at a second flip-flop and the combinational-logic output is clocked. The clock period is greater than or equal to the sum of delays (clock-to-output delay of the first flip-flop, total propagation delay through the network, and setup time of the second flip-flop). If there is significant clock skew, this calculation is wrong. In other words, the clock period might not be long enough when the clock skew is taken into account. Stated differently, in the frequency domain, the clock frequency might be too high when the clock skew is taken into account.

Clock skew can be managed by carefully routing clock signals so that they travel the same length to each chip. In fact, more sophisticated circuit-board layout software can even add length automatically to make the lengths the same. An example of this type of fix is shown in Fig. 14.2.

Some textbooks define "clock skew" as the total shift in the clock (including clock jitter discussed below) and the skew due to differing trace lengths as "spatial clock skew." In this book, the term "clock skew" refers to skew caused by spatial constraints.

Clock jitter is a measure of how imperfect a clock period is. One way to think about jitter is this: Visualize using an oscilloscope on a clock, setting it to trigger at a rising edge. Then turn the knob of the scope and view a point in time, say, several hundred clock periods later. Ideally, the clock edges should line up exactly, but they

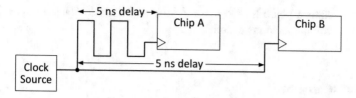

Fig. 14.2 Adding length to shorter traces to minimize skew

won't. After one hundred clock periods, some edges will be a little early and some will be a little late. One can make the measurement (i.e., measure the exact moment a rising edge occurs roughly one hundred clock periods later) numerous times and compute a mean and standard deviation. The RMS *clock jitter* is proportional to the standard deviation of the measurement.

There are two ways to consider the jitter measurement.

The first, *cycle-to-cycle jitter*, compares the period of one clock cycle to the period of the next clock cycle. The RMS cycle-to-cycle jitter is the standard deviation of this measurement made over 1000 clock cycles. The peak-to-peak cycle-to-cycle jitter is the largest value of the measurement made over 1000 cycles.

For example, consider measuring a 1 GHz clock. One clock has a period of 998 ps and the next clock has a period of 1002 ps. The difference between the two periods is 4 ps. The next clock has a period of 996 ps and so the next difference is -6 ps. The standard deviation of 1,000 "difference" measurements is the RMS cycle-to-cycle jitter and the largest difference (to be precise, the largest absolute-value difference) is the peak-to-peak cycle-to-cycle jitter.

Importantly, the cycle-to-cycle jitter is a relative measurement; it measures the jitter of each clock period compared to other clock periods.

The second, *period jitter*, compares each clock period to the "ideal" clock period. Going back to the example 1 GHz clock, the first difference, found by comparing the 998 ps clock period to the ideal 1000 ps clock period, is -2 ps, the second is 2 ps, and the third is -4 ps. The standard deviation of 10,000 of the comparison-to-ideal "difference" measurements is the RMS period jitter and the largest difference (to be precise, the largest absolute-value difference) is the peak-to-peak period jitter.

So which measurement is more useful? It depends.

In designing digital circuits, calculation of the clock period is extremely important – it constrains performance (thus there is a motivation to raise the clock frequency) and correctness (thus there is a motivation to raise the clock period). Period jitter is a measure of the jitter of a clock relative to a "perfect" version of the clock, and so is useful for measuring a clock that is being used to clock synchronous digital systems. In other words, it is useful for onboard clocks.

In modern communications, the clock and data are multiplexed on a single high-speed channel, and so the ability of the receiver to lock to an incoming signal and reconstruct the clock is essential. The cycle-to-cycle jitter expresses how "jumpy" the clock is from cycle to cycle, and so expresses how hard it is for a receiving element, such as a phase locked loop, to lock to the clock. Cycle-to-cycle jitter is useful on a communications channel.

Sources of Jitter

As you noticed, there were two types of jitter measurements, RMS and peak-to-peak. To understand when they are used, it is important to understand the sources of jitter.

Essentially everything we have discussed in the previous chapters of this book can cause clock jitter, including ground bounce, reflections, crosstalk, power supplies, EMI, and ESD. It is also caused by random processes such as thermal noise.

If the source of jitter operates on a relatively small scale and is uncorrelated to the clock, then it appears as random jitter and is measured using RMS statistics. If the source of jitter is either an occasional burst of significant energy or is correlated to the clock, then it appears as peak-to-peak jitter (either a large amount of jitter or none). By computing (and comparing) the RMS and peak-to-peak jitter, the origins of the jitter can be diagnosed. There are also formulas that combine peak-to-peak and RMS jitter. These formulas are complicated, involving an estimate of a channel's target bit-error rate, and so are outside the scope of this chapter.

Circling back through the sources, the first is random processes inside electronics, such as thermal noise. Just as there is never a channel with an infinite signal-to-noise ratio, there is never a noise-free circuit and therefore there is never a noise-free clock source.

Signal-integrity issues on the clock trace, such as poor termination, variation in geometry, and added capacitance or inductance, will cause measurable degradation. This noise is correlated to the clock.

Coupled-noise issues, such as crosstalk, EMI, and ground bounce, can also perturb clocks. These are potentially more troublesome because they arise from neighboring circuits and so may not appear in the lab. (Once again, this highlights the value of testing a system with every feature being exercised at the same time.)

Uncorrelated jitter, especially if it is relatively small and has a roughly Gaussian distribution, is called *random jitter*. Correlated jitter, or jitter that causes a large, repeatable variation in the clock period, is called *deterministic jitter*.

Clock Sources

In order to minimize clock jitter, we also have to understand the physical origin of digital clocks.

In today's technology, almost all clocks start with a *crystal*. The crystal is normally a literal quartz crystal that has a specific geometry and has been cut along very specific lines (relative to the quartz crystal lattice) in order to vibrate at a specific rate. It is very much like a tuning fork, except that it typically vibrates at a frequency of around 1–20 MHz. The vibration occurs due to the piezoelectric effect, which has to do with the fact that a crystal vibrates if you apply an electric field, and, conversely, creates an electric field if you make it vibrate.

A crystal by itself will not do much, but if it is placed in an electric circuit, the circuit and the crystal can be made to oscillate because the electric circuit locks into step with the vibrating crystal. A typical circuit is to connect an inverter across the crystal (yes, an actual inverter gate), with one end of the crystal connected to the inverter output and the other end to the inverter input. The circuit never stabilizes on a single logic value, which is desirable in this case because the whole purpose of the circuit is to oscillate. In technical terms, the circuit is *astable* – it is not stable (never reaches an equilibrium) and is not unstable (never goes off the rails). It is important to note that the voltages going in and out of the crystal are extremely small and very sensitive. Even trying to probe a crystal with an oscilloscope probe will usually throw the crystal off, and usually will stop oscillation. This is another example of how sensitive clock circuits are – they are difficult to probe directly.

A crystal is very sensitive to radiation. (In fact, it makes an excellent radiation sensor.) Hence naturally occurring radiation is one inescapable source of clock jitter. It also means that quartz crystals are unsuitable in high-radiation environments like space travel.

The combination of a crystal and some amplification or inversion circuit is called a crystal oscillator (XO). Note that the amplifier output can normally be probed, and so using an oscillator, instead of a crystal by itself, creates a circuit that can be probed and measured. It also creates a signal that is more immune to coupled noise because it is being actively driven with a significant current.

Many modern integrated circuits have a built-in amplifier, and so are designed to be connected directly to a crystal. In such circuits, there are two signals designed to be connected across the crystal, and a capacitor is connected to ground at each end of the crystal. The value of the crystal is called out in the datasheet of the integrated circuit and the capacitance value of the capacitors is called out in the datasheet of the crystal. The datasheet may also specify what type of crystal needs to be used.

One common variation of the crystal oscillator is the voltage-controlled oscillator (VCO or VCXO). This is a circuit that is designed to have a variable output frequency that is controlled (as the name implies) by a control voltage. In a typical VCO, a higher control voltage produces a higher output frequency. Some VCO's are designed for very small adjustments in frequency (parts per million) and others for very wide changes in frequency, with the range of adjustment dictated by the application.

It turns out that inexpensive quartz crystals only work well up to a few dozen megahertz. So how do modern computers generate clock frequencies in the gigahertz range? The answer is *phase-locked loops* (PLLs).

A PLL works by comparing the phase of the input clock to the PLL output and adjusting the frequency to make the phases match.

In a typical PLL, there is a circuit that performs a phase comparison (between a reference clock input and the circuit output) and produces an error voltage. The error voltage is low-pass filtered and used as an input to a VCXO. The output of the

VCXO is then divided down and compared to the input clock. (That is, the output of the VCXO is divided down and used as feedback to the error calculation.)

The dividing-down step is crucial – by dividing down the circuit and then performing the comparison, the circuit is comparing a fraction of the output clock to the input clock. Stated conversely, the output of the VCXO is actually a multiple of the input clock frequency. For example, consider a circuit where the output of the VCXO is divided by 10 and then compared to the incoming clock. Since the circuit is locking one-tenth of the output clock to the input, the output is actually ten times faster than the input.

In other words, the PLL can function as a clock multiplier, and this is the process by which a 30 MHz crystal, for example, can be used to create a 1.5 GHz clock.

There are two problems with this. First, the PLL is full of extremely sensitive analog electronics such as phase-error measurement circuits and low-pass filters. This would not be a problem except that the sensitive analog circuits are in the middle of an entire chip of rapidly switching digital circuits. Second, the PLL by its nature can multiply clock jitter. Any jitter in the input clock is magnified because the PLL is designed to lock to the input clock.

Consider a 100 ps clock jitter on a 30 MHz crystal. Normally at 30 MHz, with a clock period of 33 ns, a jitter of 100 ps is not noticeable. If the crystal is then used to generate a 1.5 GHz clock, the output potentially has 100 ps of jitter, since the output is only as accurate as the input. At 1.5 GHz, with a clock period of about 666 ps, 100 ps of jitter is much more noticeable.

This observation leads to an important layout rule – when laying out the crystal that feeds a PLL, treat the layout as if the crystal is operating at the output frequency of the PLL. Continuing with the example above, treat the 30 MHz crystal as if it is a 1.5 GHz crystal. This mental exercise will help set up the correct frame of mind for laying out the signal.

Layout to Reduce Clock Jitter

Assuming that the clock sources (i.e., the circuits inside integrated circuits that amplify or multiply the crystal output) have been properly designed, jitter comes from two basic sources. First, as the signals travel from the crystal to the integrated-circuit input, crosstalk and ground-plane return currents can inject noise into the signal. Second, the clock sources may have power-supply noise.

This leads to the layout rules that are needed.

The most effective way to prevent return current from upsetting a crystal is to remove the ground plane completely under the crystal. This is an extreme form of moating. As with connectors, this is allowable because there should not be any signals under the crystal. The capacitors to ground can be used to span the gap in the ground plane.

Integrated circuits with PLLs will almost always have separate power and ground connections for the PLL, so that the designer can minimize power-supply noise. In fact, in some designs, it may be desirable to add a series resistor and a capacitor onto the power input in order to RC filter it, or a series inductor and capacitor to LC filter it.

There is a temptation at this point to draw a figure of a typical layout, but that would be misleading. The best strategy is to study the datasheets of the integrated circuit and of the crystal and follow the layout rules there. Not only is "reading the instructions" normally the right thing to do, but also it leaves the vendor totally responsible if the design does not work.

So where do layout guidelines like these come from? You may think they arise naturally from the laws of physics, but actually they come from the school of hard knocks, as given below.

Engineer's Notebook – The Crystal Layout

Our design was like most modern ones, with one gigantic custom chip in the middle of a circuit board, with essentially all major signals going in and out of the chip. Since the circuit board was designed to be as inexpensive as possible, it had only four layers, and so the routing of the signals near the gigantic custom chip was a mess.

The lead engineer had laid out the lowly 30 MHz clock crystal as if it was carrying a slow 30 MHz signal. So it was pretty far from the chip and even traveled over a few vias.

The result was that the 1.5 Gbps SATA interface did not work. The 1.5 GHz clock driving the SATA port was too jittery. That version of the board required an extremely complicated hand rework just to get the SATA port to work, and pretty much all of the software staff was giving us frequent evil glares.

We performed a major re-layout that brought the crystal much closer and routed the signals into the custom chip with no vias. The result was a SATA port that worked.

That is the source of the rule of thumb to treat the layout as if it is at the output frequency of the PLL.

There is more to be said about the layout.

For wiring connecting a quartz crystal to an integrated circuit, keep all of the wiring on the same plane (usually the top side) and keep the crystal and integrated circuit on the same side of the board. Underneath the wiring, keep the ground plane intact. Lay out the capacitors connected to the crystal with extreme care, also keeping them on the same side.

For actively driven clock signals, such as DRAM clocks or clocks driven by oscillators, route the signals as straight as possible and with as few vias as possible. It may also help to add guard traces (see the "Crosstalk" chapter for more on this controversy).

In general, the way to minimize jitter is to maximize signal integrity. Almost every source of jitter is, in fact, bad signal integrity.

Eye Diagrams

So is there a way to visualize jitter? It turns out there is a way to visualize not only jitter but also almost every other signal-integrity effect we have studied. The way is to use an *eye diagram*.

To understand an eye diagram, consider a clocked digital signal such as the output of a rising-edge-triggered D-flip-flop. The signal only transitions at clock edges, if it transitions at all, and can have a value of 0 or 1. Over time, it might look like Fig. 14.3.

The signal shown at the top of Fig. 14.3 is a "perfect" clock – perfectly fast-rising edges, zero jitter (same clock period), etc. The signal shown at the bottom is an approximation of a real digital signal with nonzero rise and fall times, but with otherwise good signal integrity. In other words, the signal at the bottom looks like a "real" digital signal with nonzero rise and fall time propagating on a well-designed transmission line.

Notice that the rise or fall of the signal is causally connected to, and therefore shortly after, the clock edge. Specifically, in this diagram, it immediately starts to rise or fall at the clock edge. Thus, there is a half-rise-time delay from the clock edge to the edge seen in the waveform.

There is a very powerful way to visualize this even on an inexpensive oscilloscope, but to understand it, we must proceed in two steps, as shown in Fig. 14.4.

Consider all of the events that are possible at the clock edge, shown on the left side of Fig. 14.4. The signal can rise, fall, stay at a logic 1, or stay at a logic 0. In other words, at each clock edge, there is exactly one of four possible waveforms.

If an oscilloscope were set to trigger at the clock edge (the rising edge of the mythically perfect clock), and if multiple waveforms were superimposed on top of each other, the result would be the diagram on the right of Fig. 14.4. This is an eye diagram. (To be clear, the eye diagram is the diagram at the bottom, without the clock.)

The eye diagram gets its name from the fact that the waveform looks like a human eye. Good signal integrity is manifested as a clear difference between a logic 1 and a logic 0. Bad signal integrity appears as a murkier difference between 1 and 0.

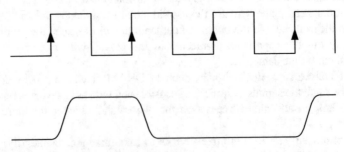

Fig. 14.3 Typical synchronous digital output

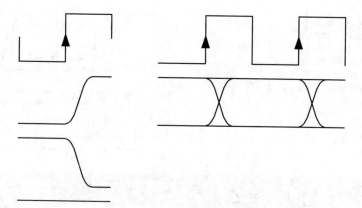

Fig. 14.4 Possible steps the signal can take (left). All possible steps superimposed (right)

Fig. 14.5 Degraded eye
diagram due to
dielectric loss

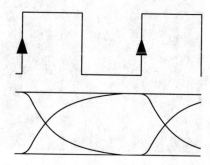

The eye diagram simultaneously captures rise and fall times, switching behavior, and clock jitter on one diagram. For example, an eye diagram for a signal after dielectric loss might look noticeably different. Consider Fig. 14.5.

The low-pass filtering of the dielectric loss causes the rising and falling edges to have a slower rise and fall time and give the edges an "RC" appearance. In the presence of more significant signal impairments, such as incorrect signal termination or an extra capacitance or inductance on the signal line, the eye diagram looks worse. For example, an extra capacitance might show up as an extra "wiggle" in the rising edge.

If the rise and fall times are too slow, the system will not work correctly. One way to express this is as a "keep out" region inside the eye diagram. Figure 14.6 shows how this works.

Figure 14.6 starts with the "good" example from Fig. 16.4. A keep-out region has been added. In order for the signal to be readily discernible by the receiver, the signal can never "touch" the dotted rectangle in the center of the eye. The bottom of Fig. 14.6 shows the degraded example from Fig. 16.5. Because the rise and fall times have slowed significantly, the signal is now too slow to be detected reliably at full speed.

Fig. 14.6 Example of an
eye diagram "keep-out"
region. (Top) Good signal
integrity avoids the keep-out
region. (Bottom) Bad signal
integrity contacts the keep-
out region

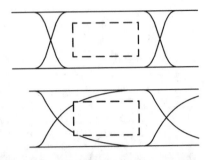

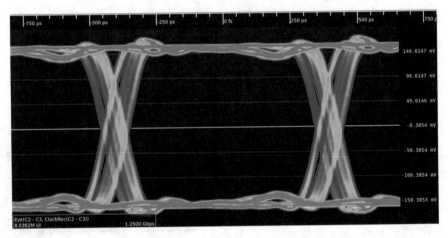

Fig. 14.7 Eye diagram example. (Source: Eye_pattern_example.png (1155 × 559) (wikimedia.
org))

Many modern standards, such as SATA and PCIe, use this keep-out-region
approach to specify what level of signal integrity is needed to be compliant with
the standard.

So how does an eye diagram show the jitter? To display jitter, a scope that is
displaying an eye diagram will superimpose thousands of signal transitions and
color-code the waveform based on how many signals are at each point. An example
of this type of eye diagram is shown in Fig. 14.7.

This shows an eye diagram from a Teledyne Lecroy oscilloscope measuring an
optical Ethernet signal (after undergoing optical-to-electrical translation). The scope
colors the waveform according to how often each signal level is encountered. So, in
this example, red signals are encountered frequently and blue signals are encountered relatively rarely.

The signal transitions are very important and this diagram shows several important properties. First, the horizontal spread in signal transitions is the actual clock
jitter. The more spread out the transition, the higher the jitter. In this example, the
transition on the left spans roughly 150 ps. Second, the transitions are not uniformly
distributed. Rather, there are roughly four peaks or "strands" visible in the

transitions. Each strand in the jitter distribution is due to some peak-disturbance type of process. In other words, the noise pattern of the channel is such that there are one of four possible major perturbances of the clock period. This is also sometimes called *deterministic jitter.* When each strand is considered separately, it appears more tightly clumped and appears to be roughly Gaussian. So each one of the four "strands" shows *random jitter.* Third, the eye diagram does not show any significant rise or fall time degradation or any other significant signal impairments.

Appendix

1. Define the following terms:

   (a) Clock skew
   (b) Clock jitter
   (c) Crystal
   (d) Astable
   (e) VCXO
   (f) PLL
   (g) Eye diagram

2. How is clock skew minimized on a circuit board layout?
3. What are some layout recommendations for a PLL or VCXO?
4. Consider the different definitions of clock jitter.

   (a) Which kind of clock jitter describes jitter from random sources on a communications channel?
   (b) Which kind of clock jitter describes jitter from a bursty, deterministic source on a circuit-board clock?

5. Alice has made some jitter measurements. On signal A, there is a Gaussian distribution of jitter. On signal B, there are two widely separated peaks in the distribution of jitter measurements.

   (a) Which signal has random jitter?
   (b) Which signal has deterministic jitter?

Chapter 15
Circuit-Board Design Process

Background and Objectives

The goal of this chapter is to step through the customary process of designing a circuit board.

When this chapter is finished, you should be able to:

- Understand the steps in the circuit-board design process
- Understand the significance of each step and the proper order of steps
- Understand how to check a design for correctness
- Have a toolkit of basic strategies for debugging boards

How Are Circuit Boards Designed?

We talk a lot in this book about circuit boards. In Chap. 2, there is a lot of discussion of how the design files get turned into actual circuit boards and how circuit boards are combined with other components to make complete assemblies. The question for this chapter is, how do the design files come into existence?

The first step is that there is some idea for a product or project. It could be an idea for something marketable, like a coffee maker, or an idea for some project, like a sensor for zebra migration, but it starts with the idea.

Depending on the nature of the idea, there may be some electronic circuitry embedded in it. This is increasingly true as computers take over the world. (In fact, I tell anyone considering whether to major in computer engineering to ask themselves a simple question – if computers are taking over the world, whose side are you on?)

You can see where this is going... So if the project or product needs electronic circuitry it needs a structure to house the circuitry and to provide the wiring and interconnection. It needs a circuit board full of circuits.

© Springer Nature Switzerland AG 2022
S. H. Russ, *Signal Integrity*, https://doi.org/10.1007/978-3-030-86927-4_15

The next step is to design the overall physical product. How big will it be? Does it need batteries? Does it require a user interface of some sort? Bad electrical and computer engineers either skip this step or do it poorly, and the result is a poorly designed product. The output of this step, whether done well or done poorly, is a concept of how the board (with the parts attached) fits into the overall product. What shape does it need? Where are connectors placed?

The output of this step can be broken down into two broad categories. The first is output from mechanical engineers (usually) – the shape of the board and the placement of connectors, displays, buttons, and mounting features. The second is output from marketing and product management – what the system needs to do, exactly. Note that for a "one-off," a board that is part of a small-scale project as opposed to a full-blown commercial product, this information may come from the actual users, such as the scientist that needs a zebra tracker.

Only at that point, after the idea for what is being designed has begun to gel, can the design of the actual circuit board begin. There is much more that can be said on the subject – for example, in a mature process, the board designers reserve the right to flag major issues that may cause the overall concept to be rethought – but that is outside the scope of this chapter.

Component Selection

The circuit-board designer starts by choosing parts. Sometimes this is easy (I want a 4.7 kilohm resistor) and sometimes it is more complicated (do I choose a capacitor with a 100-degree temperature rating or 150-degree temperature rating).

Normally the "big" components are selected first, like the processor and any specialized circuits (like FPGAs, memory, displays, analog circuits, or power supplies). This is an art, trading off features and functions versus cost, size, and power consumption.

At some point, smaller components are selected, like resistors and capacitors. This is normally more straightforward – the component values are typically dictated by other parts of the design and the package size is a function of how big (or, more accurately, how small) the final product needs to be.

The only variables on resistors are the resistor value, package size, and rated power dissipation. And they wind up being relatively straightforward.

Capacitors are more nuanced. . .

Engineer's Notebook – More About Capacitors

There are three basic types of capacitors, as illustrated in Fig. 15.1.

The first is an aluminum-electrolytic capacitor. These usually are cylindrical and have leads sticking out. They are created by rolling up foil, an "electrolyte," and

Fig. 15.1 Types of capacitors. Left to right: Aluminum-electrolytic, ceramic, and tantalum

paper insulation, hence the cylindrical shape. On the minus side, they have a lot of resistance and inductance (relatively high ESR and ESL) and the electrolyte actually dries out over time. (So they actually wear out.) On the plus side, they are available in very large capacitance values (10s of Farads) and are relatively inexpensive.

Heat, including heating from ripple current, causes them to dry out faster. Ripple current is a large concern when the capacitor is used in a switch-mode power supply because the current is basically always rippling. If one is concerned about product life (e.g., due to extreme temperatures or high ripple current), selecting an aluminum-electrolytic capacitor with a higher temperature or voltage rating will increase the lifetime. This is a place where a cheap capacitor from a shady vendor can cause significant quality issues.

One important parameter of aluminum-electrolytic capacitors is that they are *polarized* – they have a distinct positive and negative side because of the electrolyte. If they are reverse-biased (e.g., if they are soldered in backward) they overheat and can actually explode. In fact, the top of an aluminum-electrolytic capacitor usually has an X-shaped score in it so that it explodes up (into air) and not out (damaging other components). This is why old-timers (like me) never lean over a board the first time the power is applied. They almost always have at least one stripe to mark the positive or negative side. (In Fig. 15.1, the negative side of the aluminum-electrolytic capacitor is marked.)

The second is a ceramic capacitor. They are formed by interleaving sheets of metal with insulating ceramic material, and so tend to be flat, like little shoe boxes or discs. Relative to aluminum-electrolytic capacitors, they have much lower ESR and ESL and last longer. The only minus is that they are typically only available up to a few microfarads. Generally speaking, these make the best capacitors unless you need a lot of capacitance.

The third is a tantalum capacitor, so-called because they are made using tantalum in the construction of the anode. Relative to ceramic capacitors they are smaller (i.e., a 5 μF tantalum capacitor is smaller than a 5 μF ceramic capacitor) but they have significantly higher ESR. In addition, tantalum is apparently only found in politically unstable countries and so tantalum capacitors become very expensive at random (every few years).

It is important to note that all capacitors have a voltage rating. Because the goal of the manufacturer is to pack in as much capacitance as possible, the dielectric can be

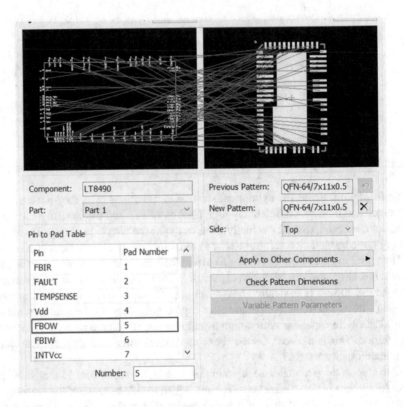

Fig. 15.2 Example of a complete component

quite thin. If too much voltage is applied, there can be an arc through the dielectric which ruins the capacitor. So the capacitor should always be chosen to have a voltage rating higher than the applied voltage of the circuit where it will be used. As explained above, designers will sometimes over-rate a capacitor to extend life.

Component selection finishes by creating a set of information for each component. Specifically, each component needs a schematic symbol and a component footprint so that the component can wind up correctly installed on the circuit board. The schematic symbol shows every "pin" (every external connection) of the component, so that it can be properly wired in the schematic. Hopefully, each pin has an intelligible name. The component footprint is the physical arrangement of the component from the point of view of the circuit board. For example, a ceramic capacitor's footprint is two rectangles, one for each end of the capacitor. The footprint is also mapped to the schematic symbol, so that the schematic and layout agree on which pin is which. An example of a component's schematic symbol and footprint is shown in Fig. 15.2.

Consider the example shown in Fig. 15.2. On the left is the component's schematic symbol and on the right is the component's surface-mount footprint. The signal highlighted in red shows both the pin on the schematic symbol and the

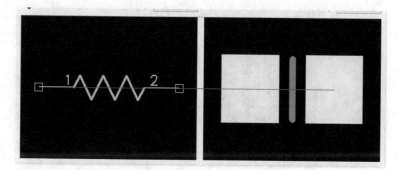

Fig. 15.3 Two views of a resistor. (Left) Schematic view. (Right) Layout view

corresponding location of the solder pad on the circuit board footprint. In this design tool, the schematic pins and surface-mount footprints are connected with blue "ratlines."

Another, more mundane, example is a resistor: If a resistor is surface-mount, for example, the footprint is two rectangles of solder because that is what sits on the board. (If the resistor is through-hole (contains wires on both sides), the footprint is two holes in the board.) An example of a resistor mapping is shown in Fig. 15.3.

This simple example shows how the circuit-board tool thinks about a resistor. On the left is the schematic view of a resistor, and on the right is the layout view. They both represent the resistor but are viewed from completely different perspectives.

To recap, the circuit-board designer creates a schematic symbol and associates the symbol with a component footprint. The association maps each symbol pin to a solderable location on the footprint.

Schematic

Once the parts are selected, the designer creates a *schematic* which is a diagram that shows all the parts and how they are connected. The schematic, for example, tells where the 4.7 kilohm resistor is connected into the rest of the circuit. An example of a schematic page is shown in Fig. 15.4. Note the component in the middle of the schematic – it is the same component shown in Fig. 15.2.

Most designs span multiple schematic pages. It is not uncommon for a schematic to have 30 pages.

The schematic may be the most important design document. It specifies the topology (interconnectedness) of the design and calls out information like what value the resistor is supposed to have. It specifies every aspect of the wiring. Which chip drives the interrupt pin? Does the sensor run off of 3.3 Volts or 5 Volts? Do I tie this address pin to power or ground? Since each component on the schematic is linked to all of the part's information, drawing the schematic is the

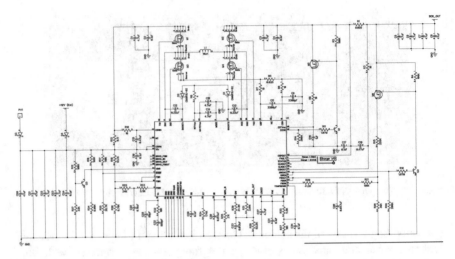

Fig. 15.4 Example of a circuit schematic page

point at which component-level decisions are made. For example, the inductor in Fig. 15.4 had to be selected to meet the current needed to be output by the design.

The schematic shown in Fig. 15.4 might seem intimidating, but keep in mind two points. First, as you gain experience schematics like this start looking easier. Second, often manufacturers of integrated circuits, such as the manufacturer of the charge controller on this schematic page, produce a *reference design* that shows how the part is normally wired in with parts around it. The reference design usually includes recommended components and formulas to calculate passive component (R and C) values.

The schematic contains reference designators, which match the markings on the circuit board. In other words, it contains information that lets you map the physical board in hand to the schematic. There are some conventions for assigning reference designators. For example, resistors start with R, capacitors start with C, diodes with D, and integrated circuits with U. In the category of resistors, each one has a unique number. Most circuit-board design tools will handle this automatically, incrementing the number each time a new resistor, for example, is added to the schematic. So the 18th resistor added to the design is R18. The important point is that "R18" refers to the schematic symbol, the part footprint on the final circuit board, and the exact type of resistor (value, package size, etc.) that goes there. The reference designators are used in all of the steps in the process of designing, procuring, and assembling circuit boards.

Many companies have part-numbering systems so that a 4.7-kilohm surface-mount resistor has a unique internal part number. If so, this information is also on the schematic.

The schematic contains important properties. One property is "do not populate" which is added to components that are not needed. Since designing a circuit board is complicated, there is often a motivation to add extra components in case they are needed later. So it is common to add an extra component and mark it "do not populate." If this is done, there is a footprint on the board for the component, so it can be added later if need be, but the component is not normally populated, so that money is not wasted to purchase it.

There is usually a back-and-forth between the schematic and the component selection. Rarely are all the components selected before schematic entry begins, and often the design decisions forced by the schematic (are you sure you don't need a protection diode?) entail changing components or selecting additional ones.

Once the schematic is complete, if the designer is smart, then a design review is held and several people go over the schematic to look for mistakes. Like every other step of the process, creating a schematic is error-prone and so this step acknowledges this fact and brings in other people to double-check the design. For example, it is very easy to get the power and ground wiring wrong, or even to forget it entirely.

There are a couple of "tricks" that might help schematic design. First, if you have a schematic that spans several pages, tie the reference designator numbering to the schematic page number. For example, if a resistor is on page 12 of the schematic, it might be numbered R1204. This enables almost anyone who is looking at a physical board to find the schematic page where the component is found. Second, use the "no-connect" symbol in the schematic software to mark pins that are left unconnected on purpose. This means that any pin that is truly unconnected is a mistake, and the schematic software can issue a warning.

Bill of Materials

Since the schematic contains every component, it is easily converted to a list of every component on the board. The schematic entry software has tools that let the designer export a parts list. This list includes not only the type of component ("10 kOhm resistor") but also its reference designator (R1204) and other important information, such as its package size (0805).

This parts list becomes the first draft of a design's *bill of materials*, a complete list of all of the components that are needed to assemble the product. If a project is time-sensitive, the parts list can become the product's first-cut bill of materials, and procurement can start ordering parts. This is a big time saver, because it lets the parts come in while the board is being laid out.

Layout

From the schematic, a *layout* is created. This is the step that creates the physical layout.

For each part in the schematic, recall that there is a *footprint*, a layout of the part as it sits on the board. There are subtleties in the footprint. For example, if the part is

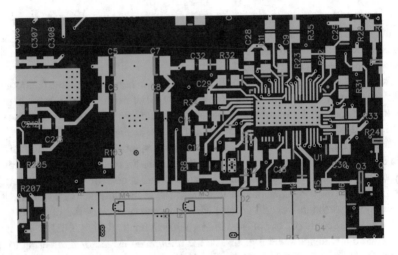

Fig. 15.5 Typical layout

a diode there needs to be some way to indicate the polarity of the diode. The layout is where the designer expresses information like which layer is used for what and where holes need to be cut into layers for vias.

To continue the example, the part footprint from Fig. 15.2 was placed into a circuit with the wiring shown in Fig. 15.4. The designer then went into the layout tool, placed each component on the board (both top and bottom), and added the wiring to connect components. The step of placing components is called *placement* (OK, that's pretty obvious) and adding the wiring is called *routing*. Once the parts have been placed, the tool will usually indicate the missing wiring with straight lines called *ratlines*. The ratlines are helpful – they show how nets are interconnected and aid the designer in placement.

Most circuit-board design packages have tools to automate placement and routing, but only use these tools carefully. They have been known to produce disastrously bad results.

An example layout (corresponding to the earlier schematic and component) is shown in Fig. 15.5.

Figure 15.5 shows how the physical circuit board both houses each component and provides the needed interconnecting wiring. Thankfully there are automated tools that enable the designer to compare the resulting layout to the original schematic, as well as to check for routine errors like shorting signals together.

Like the schematic, when the layout is finished there is another design review. Note that layout is what typically creates signal-integrity issues. The layout governs how signals are routed (laid out on the board), where the planes are, where components are, the size and shape of the circuit board, etc. Mistakes in any of these can create signal-integrity issues.

The layout includes the reference designators. Recalling the discussion in Chap. 2 about circuit boards, the wiring is found on wiring layers and the reference

Fig. 15.6 Example of
markings

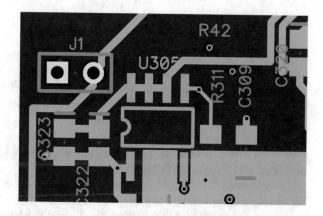

designators are found on the silk-screen layers since they need to be human-readable.
The layout is where additional reference designators can be added, like high-voltage
warnings, company logos, or the names of the designers.

Other important information can be added, like markings to indicate pin 1 or
diode polarity. These markings are essential – they are the only way that the people
responsible for assembling the components onto the circuit board know which way
to place them.

An example of a polarity marking is shown in Fig. 15.6.

In Fig. 15.6, U305 is highlighted in the layout tool. The "U305" marking is the
reference designator. Both the lettering "U305" and the rectangle inside the pins are
on the silkscreen layer. The notch in the rectangle shows the assembly technicians
where pin 1 is located. The component itself has a small dimple on top next to pin
1, and so the technicians know which way to orient it. In a modern surface-mount
line, the part is shot onto the board by a pick-and-place machine, but even so, the
technician must program the machine to shoot the part on in the correct orientation.
The "U305" label has been placed outside the part, so that it is still visible after the
part has been soldered down.

Other polarity markings are common. For larger surface-mount components,
there is often a triangle or dot next to pin 1. For diodes, the "pointy end" is marked
with a line or dot.

A final word about layout: The transition from schematic to layout to board
usually under a lot of time pressure. The rest of the company needs these boards
(e.g., software engineers need actual units in order to write the software). This is
where shortcuts get taken and signal-integrity issues are overlooked. The more
familiar you are with signal-integrity issues, the more likely you are to do it right
the first time, which is why (ultimately) you are reading this book.

Design Files

After layout, the circuit board is ready to order, and so the designer creates the files needed for the circuit-board manufacturer to create the circuit board. The factory will also need the files, so they will know which part goes where, etc. These files are a major design milestone, as everything downstream of the creation of the files will use them to assemble and test the boards.

The most common file format for the traces, planes, and silkscreen on a circuit board is Gerber, also known as RS-274X. This file format has one significant disadvantage. First, it describes everything in terms of geometric shapes (circles and rectangles). There are no net names for example. In other words, quite a bit of board information is lost. For example, a named net "SDA" runs as a circuit-board trace from U1 to U302. In the Gerber file, the net has been replaced by a series of rectangles, and the name "SDA" is completely lost. There is no designation of any component, only the geometric description of the lettering. In fact, there isn't even lettering, only the rectangles used to build up each letter. Another property of the Gerber file format is that it is a list of plain-text commands. You can theoretically edit a Gerber file using a text editor. This is both an advantage (human-readable) and a disadvantage (human-editable).

Because the Gerber format is so reductionist, other, more modern formats are often used instead if the designers want to keep the information bundled with the board. For example, ODB++ is a more modern (albeit proprietary) format.

Another piece of information is needed to fabricate a board, the locations, and sizes of all of the drilled holes. For this, a separate "drill file" is created. The file formats are more confusing but are generally called "NC Drill files." Like Gerber, NC drill files are human-readable.

The deliverable result of a complete circuit-board design process is a set of design files, ready to be sent to the board's manufacturer, to manufacture the bare boards, and sent to the assembly facility, to know where to place the parts that go on the boards. Obviously, the files are also retained for debugging.

As discussed in Chap. 16, these files are quite vulnerable. They can be edited, and malicious (extra, unwanted) components can be added to them. Because of this, design files need to be treated with care, like software source-code files or financial documents.

The Board Is Dead. Now What?

Understanding the process by which boards are designed, made, and assembled into working products helps you debug boards that don't work. When you get the first articles back from the manufacturing facility, there is an uncomfortable few days (or, alas, weeks) as the very first boards are powered up and tested for the first time. This can be tedious or terrifying, depending on how well you did your design.

The discussion begins with a board that seems completely dead...

The first thing to keep in mind is "what is the likeliest thing to go wrong." For example, it is easy to load a pick-and-place machine with the wrong component, or for the factory to leave out an important part altogether. Noting the earlier discussion of polarity markings, placing a part backward is an extremely easy mistake to make A quick visual check of the parts will help.

It is easy to make mistakes connecting power and ground to parts, so always check power and ground first. This is a logical extension of the absolutely true maxim to make sure it is plugged in first. The author of this book has skipped this step on more than one occasion, and wasted many hours relearning this painful lesson.

After checking power supplies and voltages, be sure to check the reset circuit and the clocks. These also have the ability to kill the entire board.

Assuming the board is mostly working, the next step is to go through each sub-circuit and test them. At this point, the work can be split up among multiple engineers and typically goes a lot faster.

Besides the common sense that goes into this step, there are a couple of "tricks" that will help a lot. First, be sure to hold the oscilloscope probe correctly. If you use the ground clip, you will get ringing because of the inductance of the ground connection. (This is discussed in more detail in Chap. 17.) If you use the ground clip, it is OK as long as you know the effect it will have on the measurement. Second, write down every issue you encounter. At some point, the board may have to be redesigned and, if so, it will be very good to have a detailed checklist of every change that needs to be made. It is also easy to forget something that you noticed, so write it down immediately.

Finally, an observation... Every time you go through this make mental notes and then bring them up to management later. How could you have prevented the mistakes that get made? What mistakes are made often? What could you do to make it less likely for other groups to drop the ball? This sort of thinking improves the process you work under, improves your life, and starts making you promotable.

Appendix

For the following, select circuit-board design software.

1. Choose a microprocessor, such as an ATMega 168 which is commonly used in Arduino designs. Also, select a package style. Determine the minimum amount needed to make a bootable microprocessor including the programming interface and power and ground. Consider whether you need a crystal or can use a built-in oscillator

   (a) Create a schematic symbol for the selected microprocessor and associate the symbol with a footprint corresponding to the selected package style.

(b) Determine the typical programming connector for the processor family, create a schematic symbol for the connector, and associate the correct footprint.

(c) Decide which power connection to use and add a connector for it. One commonly used connector is a USB connector, which can be used for just power and ground if need be.

(d) Choose an additional function the processor needs to perform, such as a flash-memory part for memory storage, an LCD panel, or a temperature sensor. Create a schematic symbol and footprint for the additional function.

(e) Add at least one LED (in series with an appropriately sized current-limit resistor).

(f) For resistors and capacitors decide which package style will be used, such as 0805 or through-hole.

2. Draw a schematic of the processor, programming connector, the connector for power and ground, and the extra component for the additional function.

(a) Add the appropriate capacitance for power filtering and decoupling.

(b) Consider if any pullup resistors need to be added.

3. Construct a layout of the parts.

(a) Use a four-layer board with a power layer and ground layer.

(b) Add reference designators and polarity markings as needed.

4. If there is funding available, have the board constructed, populate the parts, and write a program to operate the additional function and the LED.

For both the schematic step and layout step, use the circuit-board design software's design-rule checking to confirm that there are no errors.

Chapter 16
Circuit-Board Attacks and Security

Background and Objectives

When this chapter is finished, you should be able to:

- Understand how circuit-board attacks can occur
- Understand typical attack mechanisms and their mapping to the design process
- Identify what type of attack has occurred
- Understand methods to detect unwanted attacks

Circuit Board Recap

As we have seen, circuit boards provide both structural support and wiring inter-connection. As a result, these boards are now critically important to nearly all aspects of modern life, from cell phones to routers to factory automation. Starting with an account of an apparent circuit-board "hack," this chapter reviews the possible ways that attacks can occur and shows examples of what an attack might look like. In one example, a fully functional commercially sold microcontroller is added to a control bus on a small satellite board (the example board from Chap. 15). The chapter concludes with the steps necessary to protect circuit-board design integrity.

First Indication of Trouble

The earliest well-known publication of the malicious alteration of a PWB occurred in 2018 in Bloomberg News [1]. The article detailed how IT professionals at Apple noticed suspicious Internet traffic on their server network. After a bout of trouble-shooting, investigators discovered an extra circuit component on recently purchased

© Springer Nature Switzerland AG 2022
S. H. Russ, *Signal Integrity*, https://doi.org/10.1007/978-3-030-86927-4_16

server motherboards. The article asserted that the extra component was roughly the size of a grain of rice. The motherboards came from Supermicro, an extremely reputable vendor in the computer-server industry. The extra circuit component was allegedly able to access the board's baseboard management controller, a small supervisory processor found on many server motherboards, and use it to send data over the Internet.

The publication launched a firestorm of controversy. One security professional quoted in the article said that finding evidence of a nation-state level attack on hardware was like finding "a unicorn jumping over a rainbow." Government agencies, however, largely discounted the news, and many regard the alleged incident as discredited.

The question that is posed here is much simpler, and hopefully less controversial. How easy is it to "hack" a circuit board? The answer is that it is remarkably easy, and that, rather than being a "unicorn jumping over a rainbow," it is a very real threat that must be taken seriously. Conversely, simple steps can be taken to make the likelihood of an undetected alteration much smaller. A recently published article summarized the threat and suggested remedial steps [2].

As an aside, it is important to understand the assumptions being made in this chapter (and indeed in this book).

First, this is not considering the security of integrated circuits. While clearly an important topic, altering an integrated circuit is complicated, requiring access to the internal structure of an IC and requiring access to very specialized fabrication technology. At any rate, such possible attacks are outside the scope.

Second, this assumes that a "hack" (more accurately, "attack") is purposeful and malevolent, to gain access to information and possibly to exert unwanted control. It also assumes that the attacker is a knowledgeable electronics professional.

Third, this assumes that the simplest way to hack a circuit board is to add an unwanted component or to modify a wanted component into an unwanted one. Subtracting a component would likely cause a readily visible malfunction. Another mode of attack would be to somehow embed an unwanted circuit into the circuit board itself, but this would be quite expensive and require access to very specialized circuit-board manufacturing capabilities. This type of attack is not out of the question, but, as you will see, simply adding a component is much easier.

Other assumptions (such as how such an attack could be carried out, how difficult or expensive it would be, or who would need to be involved) depend on the type of attack and will be explained below. As will be shown, a successful attack only requires two to three skilled individuals, and the only expense is the cost of the bare boards.

Review of Circuit-Board Design, Fabrication, and Assembly

To understand how a circuit board can be altered, it is worth reviewing how they are created and put to use.

Circuit boards are a means to an end. They are used to house electronic components and provide wiring interconnection between components. As a result, they are found in nearly all electronic products. Therefore, circuit-board design normally lies at the center of product design.

The steps to design a circuit board are explained in Chap. 15, and the steps both to fabricate a circuit board and to assemble the components onto it are explained in Chap. 2. One crucial step in the process is that the circuit-board design files are sent off to other companies both the make the bare boards and to enable an assembly facility to place components on it. At this point, the files no longer enjoy the cybersecurity protection of the company where they were created. As with any files that exit a company, keeping track of the files and preventing unwanted access becomes more complicated.

Example Attacks

To illustrate the ease with which an attack can be made, a design-file attack was demonstrated in earlier published work [4]. The example is shown in Fig. 16.1.

Fig. 16.1 Before (**a**) and after (**b**) editing the Gerber file

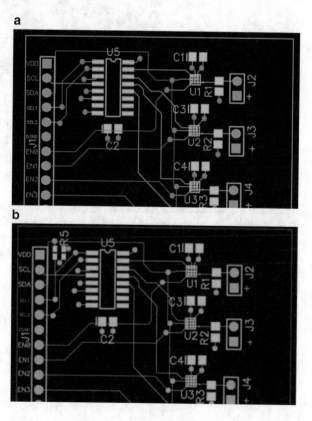

Fig. 16.2 Photo mock-up
of an altered-component
mode of attack, with the
original 2-pin footprint (**a**)
and altered 8-pin footprint
(**b**)

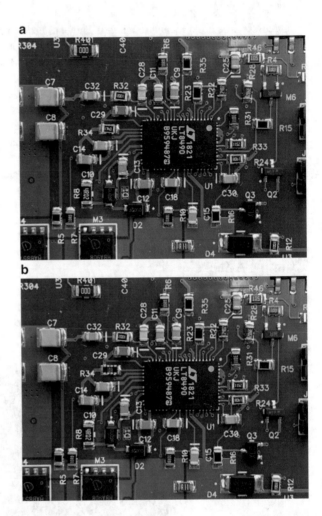

Notice the extra component on the right; it is in the upper left corner of the figure.
A four-pin footprint was added to a design's I²C bus; the footprint included power,
ground, and two I²C signals as well as the vias and wiring necessary to attach all four
signals. The component also had its own reference designator so that it would be
more difficult to spot by eye. The attack was carried out by editing a board's Gerber
file using publicly available, commercially sold Gerber-editing software.

Another example is a mock-up of an altered footprint. In this example, a two-pin
0805 surface-mount resistor is replaced with an 8-pin resistor network footprint, as
shown in Fig. 16.2.

To be clear, Fig. 16.2 was created using image-editing software, but illustrates
how difficult alterations can be to see by the naked eye. The altered component is
R34 near the center-left of the photograph.

Fig. 16.3 Original board showing processor (purple), pull-up resistors (yellow), and stacking connector (red). R504 is an 0805 component (0.08 inches by 0.05 inches) for scale

As a third example, a more complex component was added to a larger circuit board, the same board example from Chap. 15.

The goal was to add an actual, commercially sold microprocessor to an I²C bus. The microprocessor could presumably be programmed to intercept and record data or even to issue malicious commands. I²C was chosen for the proof-of-concept because it only uses two signals (named SCL and SDA) and has no native security.

To be as small as possible, an NXP LPC802 processor was selected. This processor is commercially available in a WLCSP-16 footprint containing 16 pins and measuring 1.8 mm square.

On the original circuit board, the I²C bus connects the board's processor to a lithium-ion battery over an off-board stacking connector. (In other words, the presumable reason for the attack would be to monitor or interfere with this interface.) The placement of the extra component was driven by the availability of the signals that run from the microcontroller to the stacking connector. It is important to note that a more patient hacker could hide the extra component almost anywhere, possibly burying it more skillfully in the middle of other components.

A series of pictures highlights the steps.

First, the original board is shown in Fig. 16.3 with the processor, stacking connector, and pull-up resistors highlighted. (The pull-up resistors are significant – they are required by the I²C standard and are one way that an attacker could find this bus.)

Second, the board's Gerber files were accessed in a commercially sold Gerber editor (that was legally purchased). A footprint was added just to the left of the pull-up resistors, as shown in Fig. 16.4.

To finish the design, the pins on the left side of the added component were wired to +3.3V (which is the trace shown in Fig. 16.4) and to ground (using a top-side trace, shown in Fig. 16.4, and a via to the ground plane). The two I²C connections were made on the right side of the component. One signal, SCL, was connected using a top-side trace. The other signal, SDA, was wired to a via.

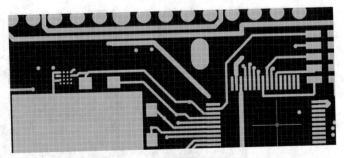

Fig. 16.4 Top-layer Gerber file shown in the Gerber editor. The footprint (a square array of 16 solder pads) has been added to the left of the two resistors. The processor and stacking connector are also shown for reference

Fig. 16.5 Added hole in ground plane for the SDA via

The ground plane needed an added hole so that the via carrying SDA to a lower signal layer was not shorted to ground. The ground-plane cut is shown in Fig. 16.5.

Figure 16.5 shows two existing vias on the left and the new via on the right. The drill hole is shown as a circle inside the cut. The odd shape is due to the fact that the hole was added by editing the traces used to construct the ground plane. Because of the Gerber format, a plane (such as a ground plane) is represented as a stacked set of wide traces. As noted in [3], this creates an odd shape when viewed afterward. With more patience, a rounder hole could be constructed.

Once the vias and top-level traces were in place, the final step was to connect SDA from the via to another location on the board. SDA already had a via, and layer 5 of the board was very sparse. So, a trace was added on layer 5 between the newly added via and the preexisting via. This is shown in Fig. 16.6.

To complete the addition of the unwanted component, the solder mask layer was edited, using NXP's recommended pattern for solder mask for a WLCSP-16 package, and a silk screen label ("U4") and a silk screen orientation marking (a dot near pin A1) were added. The final version showing the silk screen, solder mask, and top-side layer are shown in Fig. 16.7.

The added component would be difficult to notice at a glance. To provide a rough idea of the physical appearance of the board, a black square made of paper the size of the actual added component was added to the board. The complete circuit board is shown, both unmodified and modified (Fig 16.8).

Fig. 16.6 Layer-5 trace added to connect SDA to the surreptitiously added microcontroller

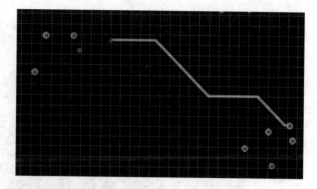

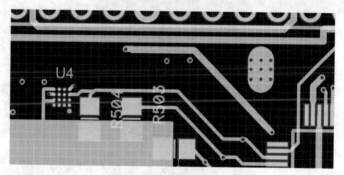

Fig. 16.7 Complete version of Gerber files with added component. The view includes the original microprocessor, pull-up resistors, and stacking connector for reference

The addition of the extra component was emulated, in Fig. 16.8, by using a piece of paper roughly the size of an actual LPC802 device. The black square piece of paper shown in the figure is slightly larger than the actual size. It can be found by locating the stacking connector and original microcontroller and then moving left to find the pull-up resistors. A "zoomed-in" view of the photo mock-up is shown in Fig. 16.9.

To recap, a footprint of a WLCSP-16 package was added to the Gerber files (and N/C drill file) of a circuit board. The footprint was selected to match a commercially sold microcontroller (specifically an NXP LPC802) and the added component was connected to power, ground, and the original board's I^{2}C bus. The modified Gerber files could then be used to create a counterfeit board with an extra, unwanted component that would be difficult to detect without careful inspection. Since the added component is a fully programmable microcontroller, and since the I^{2}C standard has no native security and permits multiple bus masters, monitoring or even altering bus traffic would be straightforward.

So how can these attacks be prevented? It helps to consider how they might be made.

Fig. 16.8 Unmodified (**a**)
and modified (**b**) circuit
board, using a black square
of paper the size of the
actual added component to
emulate the added device.
The board's actual size is
10 cm by 10 cm

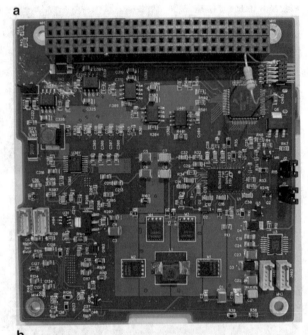

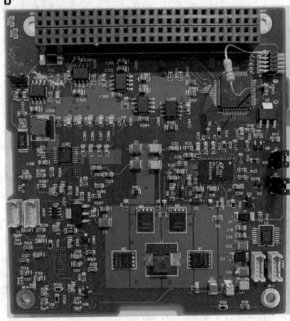

Fig. 16.9 Recreation of an actual added-component attack before (**a**) and after (**b**). The added component is a fully functional 32-bit microcontroller added to the I^2C bus

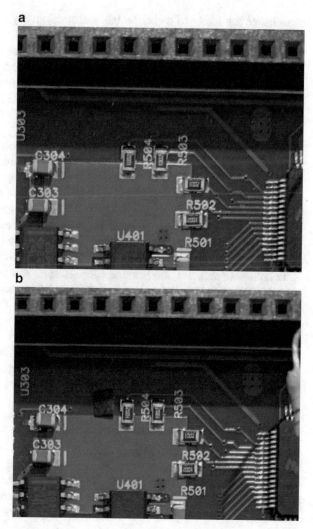

Review of Processor-Based Design

One aspect of a possible circuit-board attack is to consider where an attack might be made.

As shown above, the key to a successful attack is stealth, placing an extra component in such a way as to make it unlikely to attract notice. A corollary is that a smaller component is easier to hide than a large one, and a consequence is that the smaller number of signals involved, the easier it is to hide an extra component. Also, many modern buses use extremely high-speed signaling, and generally, such high-speed buses are difficult to modify surreptitiously – the added wiring and capacitance are likely to make the bus malfunction. So low-pin-count, low-speed buses are the most likely targets.

One common example, and the one most studied in the literature, is a serial port. Serial ports, sometimes also called craft ports, are based on the RS-232 standard and are widely used for debugging and low-level access to systems such as routers. The standard has no native security and no clearly designated bus master. One recently published article describes how a microcontroller was added to a router to give unwanted serial-port access [3].

Two buses are widely used in embedded systems and on computer motherboards, SPI and I^{2}C (also known as SMBus). SPI is a four-wire interface used in a variety of sensors and as an interface to flash memory. On a PC or server motherboard, the BIOS software is loaded over an SPI bus. The SPI standard has no native security but does have a clearly designated bus master. I^{2}C, or SMBus, is a two-wire interface used to connect sensors, such as temperature sensors, and connects a PC processor to system elements such as the power supply, clock, and fan control. The standard has no native security and natively supports multiple bus masters.

In automotive circles, the bus that controls most automobiles is a two-wire bus named CAN (Controller Area Network). Besides having no native security, the CAN bus has no centralized controller or bus master.

Reviewing the four buses, all of them are commonly supported by low-cost, low-pin-count microcontrollers.

On PC motherboards, there is another bus called the LPC or Low Pin Count bus. This bus is normally used to attach a processor to a "super I/O chip" that supports serial ports and other basic motherboard functions. The LPC bus uses seven mandatory signals and up to six optional signals.

The LPC bus normally connects a processor to a baseboard management controller (BMC). The BMC is often found on servers and enables remote access for maintenance purposes. The BMC is designed to boot up separately and has its own Ethernet port, so that even a server with no functioning operating system can be powered up and managed via its BMC. Many motherboards have a trusted platform module (TPM) which is a chip designed to provide security functions. This module is typically connected using an LPC bus, or can be connected via an equally hackable SPI bus.

To recap, there are at least five buses found in PC motherboards and embedded systems that have relatively low speed (10's of MHz clock speed or lower), low pin counts (7 pins or less), and no native security, and are therefore vulnerable to attack. These buses include RS-232 (also known as serial or UART), SPI, I^{2}C (also known as SMBus), CAN, and LPC. This inventory is important to note because it provides a good starting point for checking a suspect board.

It should be noted that it is possible to add layers of security on top of the native standards. For example, BIOS code can be protected by code signing, and the trusted platform module operating over an LPC bus has its own cryptographic protection techniques. Nevertheless, the low-level buses themselves are insecure.

Taxonomy of Attacks

Previously published work has explained that it is possible to construct a taxonomy of possible attacks [4]. The taxonomy is based on the places where an attack can be made and the manner of the attack.

First, there is a taxonomy based on the point of attack.

A *schematic attack* is one in which the circuit-board schematic file is maliciously altered. This would represent the most serious attack, as the schematic represents the entire electronic design at a symbolic level. A person with access to a schematic would have access to all aspects of the design, including which components were used and how they were interconnected. Detecting a schematic attack might be difficult, as an attacker would have a nearly unlimited number of ways to hide an unwanted component. Conversely, mounting a successful schematic attack would be difficult. An attacker would need access to an actual schematic, and therefore would either have to be an insider (malicious employee) or would have to launch a successful cyberattack. The added details would have to evade the notice of other employees, including those involved in design reviews.

A *layout attack* is one in which the circuit-board layout file is maliciously altered. One symptom of a layout attack should be clear – such a board would not match its schematic. A layout tool has powerful capabilities because it has access to the board's netlist, including important information such as signal names, pin numbers, interconnection information, and package footprints. In other words, adding a new component or modifying an existing one is easily done – it is the actual purpose the software was designed to carry out. Detecting an attack involves a careful comparison of the circuit board to its original schematic and, unless it has been corrupted, its bill of materials.

A *design file attack* is one in which the design file, such as the Gerber file, is maliciously altered. This is very similar to a layout attack in that the attacker has access to the physical layout of the board. Unlike a layout attack, however, the attacker does not necessarily know the names of signals or the identities of components, such as integrated circuits or resistor values. Conversely, with access to a complete set of Gerber files, it is not terribly difficult for an experienced designer to figure out the basics of a circuit board. For example, most designs have a single processor or microcontroller which is a strikingly large component on a design. They interface with peripherals, such as sensors or the flash memory needed for BIOS, over interfaces like SPI or I^2C. Some interfaces, like I^2C and CAN, usually contain pull-up resistors or terminating resistors that make identification possible. As with a layout attack, detection involves a careful comparison of the circuit board to its original schematic and its bill of materials. The extra component on the circuit board in Chap. 15 was a design file attack.

A *rework attack* is one in which a part is added onto a board by hand after manufacturing. This type of attack may be easier than a design-file attack because the attacker has access to a fully assembled board and can therefore identify components and pinouts. Conversely, detecting this type of attack is usually straightforward because components that are added by hand are often easy to spot. The attacker normally relies on the fact that boards inside fully assembled units are rarely examined. This was the type of attack mounted on a server serial port [3].

The first three types of attack involve altering the design of the circuit board itself. Stated differently, they entail the creation of a maliciously altered counterfeit circuit board. Mounting a successful attack involves three more steps, inserting the counterfeit board into the supply chain, populating the board with the altered component, and shipping the board to desired targets.

Inserting counterfeit boards into the supply chain may be straightforward, depending on the security of shipments from the board's fabricator to the factory where assembly occurs. All aspects of this chain have to be considered, including the receiving area of the assembly facility.

Populating the board with an altered component could be simple or complicated. A simple way to populate the board is to add it in the assembly facility's repair area. It is quite normal for newly fabricated boards to fail manufacturing testing and require minor rework. Adding a component in repair would be extremely simple, merely requiring one employee to bring in almost-microscopic surface-mount components and add them using the tools already found at the repair station. A much more complicated way would be to add the component to the assembly process. This entails adding the part to the feeder system and programming a pick-and-place machine to add the component. This would either be nearly impossible or require the cooperation of the assembly facility's management.

Shipping the boards with altered components could either be done at random, relying on adding parts to a fraction of assemblies, or could be targeted. The targeting would only require one employee that knows which assemblies have altered components and which targets are desirable. In other words, it would only require the action of one employee in the shipping area.

Note that a rework attack requires much less planning – an attacker simply obtains a system and adds malicious components to it by hand. Such an attacker could then target specific customers with relative ease.

A second aspect of the taxonomy is the mode of attack; there are two modes of attack.

The first is to change a component's footprint to add extra pins, known as *"altered component mode."* For example, a 2-pin footprint for a resistor could be replaced by an 8-pin footprint for a resistor network. The six extra pins could be used to supply power and ground to a small microcontroller with an SPI, CAN, or I^2C interface. This was illustrated in Fig. 16.2.

The second is to add a new component to the board's design. This is called *"added component mode."* It is normal for assembled circuit boards to contain population options, such as extra circuitry used in debugging and development and not in production. Hence it is normal for assembled circuit boards to contain unpopulated footprints.

To fool an inspector, an added component would probably also need a reference designator. A reference designator is the human-readable text marking found next to each component. A component without a reference designator is automatically a

Table 16.1 Taxonomy of attack locations

Location of Attack	Summary	Detection Strategy
Schematic Attack	Design schematic is altered	Double-check the schematic carefully for aberrations. Compare physical board to a trustworthy version of the schematic.
Layout Attack	Design layout is altered	Compare physical board to schematic and bill of materials
Design File Attack	Design file is altered	Compare physical board to original layout file, schematic, and bill of materials
Rework Attack	Component is added by hand after board is assembled	Look for hand-added components

Table 16.2 Modes of attack

Mode of Attack	Summary	
Altered Component Mode	Footprint of legitimate component is altered	Compare every footprint to that found in a trusted schematic
Added-Component Mode	A new component is added to the board	Compare every footprint to that found in a trusted layout

mistake, either unintentional or purposeful, because the designators are used throughout the design process to enable correct assembly. Therefore, one relatively simple way to check a board is to check for missing or incorrect reference designators.

The taxonomy can be summarized in the following two tables (Tables 16.1 and 16.2).

The likely success of a schematic, layout, or design-file attack lies in the fact that it is difficult to compare an assembled board to (in increasing order of difficulty) the design file, layout, or schematic. The board's design can be altered after the designer is finished with it but before it is fabricated. There is little to connect a fabricated board in hand back to the file that was used to create it, and the only way to be certain is to do some form of manual comparison. Stated differently, the design of circuit boards is not tamper-evident.

Checking for an attack involves comparison, first, of an actual bare board to a trusted design file and, second, of an actual assembled board to a trusted set of design files (including bill of materials). It is important to check both which components are on the board (to spot added ones) and the footprint of each component (to spot altered footprints).

If someone detects an attack, there are some simple questions that can shed light on the nature of the attack. Is the attack the result of an added or altered component? Does the component have a reference designator? Is the unwanted component found on every board or only on a subset of boards? X-rays of the power and ground plane (to see if holes are not round) and careful checking of logos added by the circuit-board manufacturer may yield additional clues.

Given the relative ease of editing the design files and relative difficulty of smuggling counterfeit components into a board-assembly facility, the likeliest attack entails altering the design file and then producing entire assemblies using the original bill of materials. If this attack is carried out, one symptom is that only a fraction of boards will have the added-component footprint.

Conclusion

Circuit-board security needs to be taken very seriously. As shown here, it is a remarkably simple matter to edit a board's design files and add extra components. Other published work has shown that it is also simple to add extra components after a board has been assembled. Companies need to ensure that the companies that fabricate and assemble circuit boards have competent cyber security and are trustworthy. Extra vigilance, such as checking for unknown reference designators and footprints, can also be carried out.

Appendix

1. Consider an I^2C bus, which requires 2 signals named SCL and SDA.

   (a) How many pins would the device need to attach to an I^2C bus? The device needs to connect not only to SCL and SDA but also to power and ground.
   (b) Bob just received a newly designed board. What would you tell him to do to check the I^2C bus and make sure an extra component was not added to it?

2. Alice has found an extra component on a circuit board. The component was soldered on by hand and its pins connect to the SPI bus that runs between an AMD processor and the flash memory that holds the BIOS (basic operating instructions).

   (a) At what point was the part likely added – in design, at the facility that created the bare circuit board, in the assembly facility where components were soldered down in a surface-mount process, or in a warehouse that stocked the fully assembled board?
   (b) What is the likelihood that a large percentage of the boards have been attacked in this way?
   (c) Which type of attack is this?
   (d) Which mode of attack is this, altered-component or added-component?

3. Bob has found an extra component on a circuit board. The component is soldered onto a complete, normal part footprint. The part footprint was not in the board's original schematic or layout, and the part's reference designator is not in the board's original bill of materials.

(a) At what point was the footprint likely added – in design or after the design files were sent to the facility that created the bare board?

(b) What steps would you recommend that Bob take to make sure that the facility that created the bare board did not add the component?

(c) Since the original schematic and layout did not contain this component, which type of attack is this?

(d) Which mode of attack is this, altered-component or added-component?

References

1. J. Robertson, M. Riley, The big hack: How China used a tiny chip to infiltrate U.S. companies. *Bloomberg Businessweek*, Oct. 4, (2018)
2. S.H. Russ, J. Gatlin, Ways to hack a printed circuit board: PCB production is an underappreciated vulnerability in the global supply chain. IEEE Spectrum **57**(9), 38–43 (2020)
3. A. Greenberg, Planting Tiny Spy Chips in Hardware Can Cost as Little as $200, *Wired*, Oct. 10, 2019. Accessible via https://www.wired.com/story/plant-spy-chips-hardware-supermicro-cheap-proof-of-concept/
4. S.H. Russ, Techniques to Thwart Surreptitiously Altered PCBs, in *2020 IEEE Physical Assurance and Inspection of Electronics (PAINE)*, (Washington, DC, 2020), pp. 1–4

Chapter 17
Testing, Debugging, DFX, and Quality Management

Background and Objectives

If you apply everything in this book to a design, chances are it still may not work correctly the first time. And then what about the 1000,000th time? How can we make sure our designs work and can be mass-produced? When this chapter is finished, you should be able to:

- Demonstrate the correct way to hold a scope probe
- Explain what the "X" in DFX stands for
- Address design issues like manufacturability, testability, procurability, and repairability
- Understand the different components of cost, price, and margin

Testing a Board: The Oscilloscope

Every computer and electrical engineer should know how to use an oscilloscope. It is by far the most useful piece of test equipment you can use or own. But it has surprising limitations which most engineers don't know about.

The first limitation is that the scope probe will produce incorrect results almost every time. Unless you know the trick. Consider two ways of holding a scope probe as shown in Fig. 17.1.

On the top of Fig. 17.1 is the way that all of us were taught to put in a scope probe. The probe has a ground clip that connects to ground, and either a point or a "witch's hat" (with a hook in it) are used to connect to the signal. If one probes a signal this way, and if the signal has frequency content over about 20 MHz, one will see significant ringing – but it may not be from the signal. Very likely the ringing is from the inductance of the loop formed by the ground clip. There is a surprisingly simple solution to this – pull off the witch's hat. (I had no idea it would pull off until I

© Springer Nature Switzerland AG 2022
S. H. Russ, *Signal Integrity*, https://doi.org/10.1007/978-3-030-86927-4_17

"Normal" probe with "witch's hat" cover and ground clip

Probe with "witch's hat" removed. An X-acto® knife is connecting the side to ground and the point is on the signal under test.

Fig. 17.1 Two ways to hold a scope probe. Top: Old-fashioned, conventional way. Bottom: The correct way

tried it.) The probe now has a sharp point and a long, skinny metallic part. The point is for probing the signal under test and the long, skinny part is the ground. The ground should be connected as close to the signal as possible using as short a connection as possible. An X-acto® knife will work; so will a paperclip. (I discovered the paperclip trick myself in a moment of desperation.)

The good news is that you don't have to do this every time. Once you use the no-hat method and make sure that the ringing is just an artifact of the probe, it is perfectly OK to keep using the probe. The time measurements are still accurate, and you can just ignore the ringing.

Finally, there are fancy scope probes, like one with two small prongs (one for the signal and one for ground). These can be specially ordered if your company has the money and can be used to probe up to very high signal frequencies.

The second limitation is one you already knew, but the consequences are surprising. Your scope has a finite bandwidth. In fact, it is usually printed right there on the side of the scope.

For example, an inexpensive oscilloscope might be listed as a "40 MHz Oscilloscope." What does this mean? It means that the input circuitry of the scope imparts 3 dB of loss at 40 MHz, and higher at higher frequencies.

If you recall from Chap. 1, we developed the notion of knee frequency (f_{knee}) and showed that $f_{knee} = 1/2t_r$ (where t_r is rise time). The rise time of a scope is similar, although slightly different since it is a 3-dB frequency:

$$t_r \approx 0.4/f_{scope} \qquad (17.1)$$

where f_{scope} is the bandlimit of the scope. So for a 40 MHz scope, $t_r = 0.4/$ 40 MHz $= 10$ ns. In other words, a 40 MHz scope creates a 10 ns rise time adder to the incoming signal.

Example 17.1
Consider a SATA signal, to be probed with two different scopes. The SATA signal has a rise time of 125 ps.

(a) What rise time is measured on a scope with 200 MHz bandwidth?

$t_r = 0.4/200$ MHz $= 2$ ns $= 2000$ ps. Measured rise time $= \sqrt{2000^2 + 125^2} = 2003$ ps $= 2$ ns. In other words, the SATA signal appears on the scope to have a rise time of 2 ns, which is completely wrong.

(b) What rise time is measured on a scope with 4 GHz of bandwidth?

$t_r = 0.4/4$ GHz $= 0.1$ ns $= 100$ ps. Measured rise time $= \sqrt{100^2 + 125^2} = 160$ ps. This is much closer to reality, but still a little off.

So if a company needs engineers to measure fast signals, the company needs to pay for a nicer scope.

The third limitation of oscilloscopes is that the scope probe can add quite a bit of capacitive loading, normally in the 10–100 pF range. If this is an issue (and it can be for very sensitive circuits, like radio circuits), then an active probe (a probe with a built-in amplifier) can have much lower capacitance.

Qualifying a Design for Mass Production

The first step to debugging a board is to bring it up to the point where it can run the software and send diagnostic messages. This is outlined in Chap. 15.

The next step is to exercise every aspect of the board under controlled conditions to make sure it actually works. The conditions should duplicate the most extreme the unit is designed to operate in. For example, a radio for a car has to operate at higher (and lower) temperatures than a radio for a home.

This process is called *design qualification testing* or *design verification testing*. The specifications for the product are turned into a detailed list of tests that verify the product meets every single specification. The specifications might include temperature, input power voltage, etc. They may include mechanical aspects such as drop height without breaking or random aspects such as which household cleaners can be used on it. The test plan incorporates all of these specifications.

As testing proceeds, any failed tests result in some sort of redesign. As with all debugging, this needs to be written down to make sure all of the failures are incorporated into useful feedback for product improvement. A mature product-design process starts this testing early, and starts with the most demanding tests first so that there is time for a redesign.

After iterating and getting all of the qualification tests to pass, the product is ready for mass production, at least from an engineering standpoint.

Transitioning to Mass Production

Say, for the sake of discussion, a product makes it through design qualification testing with flying colors. How can the designers make sure that unskilled laborers can make 1000,000 of them? This is an important question: If the laborers cannot make the product, the company that designed it will go broke.

The transition to mass production can be considered from two different perspectives, documentation and design.

From a documentation perspective, the factory will need to know essentially all of the design information, like the circuit-board layout and the bill of materials (which maps components to places on the board).

The bill of materials calls out components, and each component has an approved vendor list. The design team typically calls out which manufacturers are approved for each component, and performs qualifications (and reads datasheets) to set this up.

It is common for one design to have several different bills of material. For example, one version may have more memory than another. The versions use the same circuit board, but different memory parts are selected depending on which version is being produced. Another common example is that there may be special debug headers or features for developers which are removed in mass production.

Other significant design information might be test points (more on which in a moment) or testing strategies, repair strategies, and a plan for the mechanical assembly of the complete unit.

This documentation highlights the second perspective, that of the design itself. It should be clear that it is quite simple to design a perfect product that is nearly completely incapable of being manufactured.

This is the role of DFX, which is short for Design-for-X. To be able to be mass produced in volume, a product has to be . . .

Manufacturable: The product is easy to assemble, and the possibility of incorrect assembly is minimized.
Testable: It should be possible to test the product completely and quickly.
Reliable: It should work reliably and work for the length of its expected lifetime.
Repairable: It should be simple to repair the product.

Procurable: It should be possible to procure all the parts needed for mass production, especially if there is a surge in demand.

Disposable: By disposable in this context, it means that the product can be safely disposed of when it reaches the end of its useful life.

Every one of these is a laudable design goal (design for manufacturability or DFM, design for testability or DFT, etc.) and so they are rolled up into "Design-for-X" or DFX.

Consider design for manufacturability: Besides the mechanical assembly, the designer should consider the assembly of the circuit board itself. For example, through-hole parts have leads that stick down, so they need to be kept away from surface-mount parts so that their insertion does not damage surface-mount parts. Parts need to be kept away from mounting holes so that the humans who assemble the product are less likely to damage parts if the screwdriver slips. Can cards or boards or connectors be put in backward or inserted off by one pin or by one row of pins? Does the assembly technician have to be a contortionist to get the last screw in? These are the types of design issues that can arise.

Manufacturing facilities have elaborate systems for evaluating their accuracy and the overall quality of the product. Their main goal is to make exactly the same product every time.

So why is sameness so important?

First, it means that you can lot-sample products and be confident that all of the products are good.

Think of it this way: Most design engineering teams are happy if all of the units in a lot sample pass all of the tests. But this is not sufficient. What confidence do you have that all of the units that were not tested will pass? This is the real purpose of these calculations – statistical confidence that 100% of the products are good even though only a lot sample is tested.

(As an important aside, every time you bring a new product into a factory, the factory is not the same. This means that you are a villain when you set foot in the door. This is the primary cause of friction between design teams and manufacturing facilities. Overcoming the friction is best accomplished by you, the designer, sitting down and understanding the constraints the factory works under.)

Second, it shows that the process is repeatable and therefore more likely to remain correct over time.

The industry has a system for quantifying "sameness." The process is, first, to make a measurement on a lot sample (a sample of products from one production lot) and compute a mean and standard deviation on the measurement. Note that this process is particularly effective if a test is difficult, expensive, or destructive to run – it enables testing of an entire lot by only sampling a fraction of the lot. Second, compute the number of standard deviations the mean is from the upper and lower test limits. Consider a histogram of test results shown in Fig. 17.2.

Armed with like that in Fig. 17.2, a statistical parameter C_{pk} is calculated:

Fig. 17.2 Example of a
measured parameter

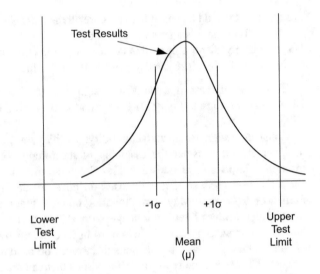

$$Z_1 \frac{(\text{UpperLimit} - \text{Mean})}{3\sigma} \qquad (17.2)$$

$$Z_2 \frac{(\text{Mean} - \text{LowerLimit})}{3\sigma} \qquad (17.3)$$

$$C_{pk} \equiv \min(Z_1, Z_2) \qquad (17.4)$$

C_{pk}, then, is the number of "3σ's" that the mean is away from the edge. What is special about 3σ? In a normal (Gaussian) distribution, an interval of $\pm 3\sigma$ contains 99.7% of the sample population, so only 3 units out of 1000 are outside the interval. A process with a C_{pk} of 1 will have 3 units out of 1000 outside the acceptable range for the measured parameter. The goal is to have all C_{pk}'s well above 1.

Example 17.2
A lot of 10,000 units has been produced. On a sample of 25 units, the video signal-to-noise ratio (VSNR) was measured. Over the sample of 25, the mean was 34.6 dB and the standard deviation was 0.4 dB. The upper limit of VSNR is 35.8 dB and the lower limit is 32.0 dB. All of the units in the lot sample were within specification.

(a) Calculate the C_{pk}.

$Z_1 = (35.8 - 34.6)/(3 * 0.4) = 1.00$
$Z_2 = (34.6 - 32.0)/(3 * 0.4) = 2.17$
$C_{pk} = 1.00$

(b) Estimate the number of failing units in the batch of 10,000 assuming a normal (Gaussian) distribution.

Since $C_{pk} = 1.00$, the mean is exactly 3σ from the upper test limit. So 99.7% of the units are inside the limit and 0.3% are outside. 0.3% * 10,000 units = 30 defective units in the batch of 10,000.

(c) Is this acceptable?

Is 30 units defective out of 10,000 OK? If these were medical devices, for example, the answer is clearly "no." In most cases, this level of quality would not be acceptable.

There are numerous other aspects of design for manufacturability. One crucial aspect is ease of assembly. For example, how many different types of screws are needed? Are all screws inserted from the same direction or do you have to approach assembly from multiple sides? Can hand-inserted parts or connectors be put in backward or wrong?

Engineer's Notebook: Confusing Connectors
One of our set-top designs had an optical audio connector next to an Ethernet port. We discovered that many of the Ethernet ports were coming out of production with bent pins. Our conclusion was that the factory test personnel were jamming the optical connector into the Ethernet port by mistake. The two ports do look alike, and were located right next to each other. This product never made it into mass production, but if it did, I suspect a great deal of units would have come back for repair – consumers would probably also be confused by these connectors.

On other designs, the USB and SATA ports were located next to each other, and often test personnel mistook the two. This crunched the small circuit board inside the connector. (In fact, you can insert a USB connector upside-down if you try hard enough. Of course, you break it when you do it.)

Finally, one group designed a cable set-top box for the European market. There was a large hideous television connector for the European market called a SCART connector which has a large metal connector and a thick cable that comes off of it at a right angle. The thick cable went right over the HDMI port, and many units came back with the HDMI connector broken. Because of the way it was routed, the SCART cable pressed down on the HDMI cable, and the relatively weaker HDMI connector broke.

In all of these cases, we (the product designer) did not take into account how the product was actually going to be used. (Remember that the test personnel in the factory are actually using the product the same way that consumers will – it's just that factory workers will tend to install all of the cables, whereas consumers might only install a few.) This type of common sense is essential for good design.

Consider design for testability: Actually, one could write an entire book just on this subject, but some basic thoughts will help you get most of it.

How are electronic products tested? The first observation is that it is generally best to test as soon as possible in the process. This keeps defective products from running around the factory and lets everyone know, at least approximately, the source of the problem. Once a board is assembled (has all of its components placed and soldered on), it is typically tested by itself. In most modern designs this is almost all of the electronics. Since it is a board by itself with no casing, a fixture is needed to hold the board and supply power.

Many of the circuits on the board can be tested functionally. If the board produces a video or audio output that is correct, then the components that go from the IC that generates the output to the connector are probably all good. Memory is a classic example of a circuit that can be tested functionally.

Some circuits cannot be tested functionally, however. A classic example is a pull-up resistor (a resistor that puts a signal in a known state if no outputs are driving the signal). Many circuits will work just fine without the pull-up resistor, but that is sheer dumb luck. Often the luck runs out after you have built and shipped a few thousand units.

For circuits that fail intermittently, more detailed testing is needed.

A common approach is to build a fixture with dozens (or even hundreds) of pins that contact the board. The fixture is called a "bed of nails" tester. Each pin contacts a point on the board called a "test point." The nice part about this approach is that, while the fixture costs money, you only have to buy it once and the test points are essentially free, just pads of copper on the bottom side of the board. With a test point next to a pull-up resistor, the bed of nails tester can measure the resistance and confirm the resistor is there.

The alternative, which is also commonly used, is to install a camera and perform automated optical inspection. The advantage of optical inspection is that you don't need to build a test fixture for each board. The disadvantage is that you cannot confirm a component's value, only its presence.

Consider design for repair: This turns out to overlap with design for manufacturability and testability. How easily is the product disassembled? Once disassembled, how easily is it tested?

Consider design for procurement: This one is very important. How will your company buy large quantities of every single component? The goal of procurement is "just in time" (or JIT) in which the components arrive just as they are needed for assembly into units. The procurement group will therefore prefer components and vendors with short lead times – parts that can be ordered quickly – so that the factory can be ready if there is a surge in demand. Unfortunately, procurement will also want to bring in the least expensive vendors, and so engineering has to scrutinize their selections. Even seemingly routine parts, like capacitors, can have poor quality and cause product failures.

Engineer's Notebook: Factory Fire

Some components are only manufactured by one company. In the case of very specialized parts, like large integrated circuits and microprocessors, this is quite common.

Our company had one of these *sole-sourced parts* (parts only manufactured by a single vendor) in almost all of the units in mass production. One day the factory that made the part burned down (literally, true story). The company then announced that they were just not going to make the part anymore.

The hard lesson is this: If you (engineering team member) want to select a sole-sourced part, make sure the company that manufactures it has a *business continuity plan*. That is, a euphemism for a plan for what to do if their factory shuts down.

Appendix

1. A coworker has probed a signal on a circuit board and is freaking out because there is a great deal of ringing. What needs to be checked before anyone freaks out?
2. A coworker is probing a signal on a circuit board. The signal shows a 3.2 ns rise time.

   (a) The oscilloscope of the coworker has a 200 MHz bandwidth. What is your best estimate of the actual rise time of the signal?
   (b) What would be the measured rise time if the coworker used a 1 GHz oscilloscope instead?

3. What scope bandwidth is needed to probe a signal with a rise time of 1 ns?
4. A lot of 100,000 units of an MP3 player has been produced and shipped. There are numerous customer complaints about the product. 100 of the units are brought in and the audio signal to noise ratio (SNR) is measured. The measurement is found to have a mean of 52.3 dB and a standard deviation of 3.3 dB. The product is supposed to have an audio SNR of 45 dB.

   (a) What is the C_{pk} of the product?
   (b) What does the Cpk result mean?

5. Consider DFX.

   (a) What does DFX stand for?
   (b) Name three examples of "X"
   (c) Which "design-for" goal attempts to minimize the time it takes to fully test each board?
   (d) Which "design-for" goal attempts to make products easier to dispose of at the end of the product's life? What are the environmental and societal benefits of this goal?
   (e) Why is design for manufacturability important?

Chapter 18
Commercial and Legal Implications, Project Management, and Risk Mitigation

Background and Objectives

This chapter introduces you to some more of the "other aspects" of design. While not necessarily technical aspects, they are essential for good design, and knowing about them will help you be a better designer. More importantly, it will help you and your coworkers produce a design that is successful. This information will also help you decide whether you eventually want to go into engineering management and therefore help you make some long-term career decisions. When this chapter is finished, you should be able to:

- Describe some of the requirements a design must meet in order to be legal for sale
- Describe some of the commercial aspects of design, such as profit margin
- Understand the basics of project management and scheduling
- Develop an understanding of the design process and how to improve it
- List sources of technical risk and some strategies to manage the risk
- Understand possible engineering career paths and start planning for them

Legal Aspects of Design

We have touched on this briefly in other chapters, but what requirements must a product meet in order to be legal for sale? While the answer clearly varies greatly depending on the product, the industry it is being sold to, the customer, etc., there are some elements of product design that are necessary for all products.

The first and most important is safety certification. This is both a legal requirement and, clearly, an ethical one. Fortunately, there is a well-documented process to follow.

In the US, the requirement for safety certification is actually driven by workplace legislation which is enforced by OSHA, the Occupational Safety and Health

Administration. The requirement is that all the electric items in a workplace be certified by a Nationally Recognized Test Lab (NRTL). One example is Underwriters Laboratory (UL). (Others can be found on the OSHA website.)

Underwriters Laboratory was started by insurance underwriters to reduce the number of fire insurance claims. To this day, much of UL requirements center around devices not bursting into flame.

To make a long story short, a product must be designed to UL standard 1950 or international IEC standard 950. (The two are harmonized, fortunately.) To meet the standard, the test lab has to certify the design (circuit board layout, choice of materials, etc.), has to test an actual sample of the design, and has to certify the manufacturing process.

The first two are obvious, but why the third? It is very simple – they want to make sure that not only that the first product is safe, but also that all of the products that are manufactured are safe. It safeguards against fraud (a company built a safe product for them to test and then built a bunch of cheap ones to sell) and against negligence (the factory starts to take shortcuts). For example, in order to have a certified process, 100% of products with a power cord have to be *hi-pot tested*. (High voltage is applied to the power input and the leakage current is measured.) Once everything has been certified, including the manufacturing process, the product is said to be *listed*.

Backing up to the design, the UL/IEC standard calls out a series of guidelines that have to be followed. For example, AC power wires have to be a certain color and have to be tied down two separate ways. Needless to say, the details are outside the scope of this book and need to be consulted separately.

One of the requirements is that only recognized components can be used in the power supply; these components have their own safety certification process. In the US, components are labeled "UL Recognized" and have a backward "RU" on them. The components cannot be UL listed, because as noted above the listing only applies to complete products and manufacturing processes, but the recognition means that the component is safe for use in a properly designed (and properly manufactured) power supply.

The test lab (and others) can provide training services to review the standards, and the test lab can come out and go over your design early in the process. So this turns out to be straightforward.

It is interesting to note the contrast between electrical and computer engineering on one hand and, say, civil engineering on the other. (In fact, many civil engineers misunderstand this.) In electrical and computer engineering, the *product* is certified, not the engineer and not the design. This is because it is a mass-produced product, can be destructively tested, and has been certified to meet a list of well-recognized standards.

The second certification is radiated emissions testing, at least for any electrical product with a clock frequency above a few dozen kHz. This is discussed in detail in the chapter on EMI/EMC.

If the product is to be sold in Europe, additional testing is needed in order to get the CE mark, including ESD testing, susceptibility, and dips and interrupts. So the

CE mark is the combination of safety certification, radiated-emissions certification, and these extra tests. ESD design and testing are discussed in Chap. 13.

It needs to be noted that there may be substantially more certification required, depending on the industry the product is designed for. For example, the nuclear power industry has extensive software-development certification requirements. As noted in the sections on EMI/EMC and ESD, different types of products might transmit radio-frequency energy or be connected to the public switched telephone network, and so different guidelines may apply.

A third set of certification has to do with intellectual property. If a product uses or adopts someone else's technology, it must be licensed, and a product should not infringe another company's patents. (Patents cover inventions such as innovative manufacturing processes, materials, or methods.)

In the case of most intellectual property, the role of engineering is to identify expertise that is needed from outside the company and bring it to the attention of the legal department to make sure it is correctly licensed. For example, when purchasing a compiler, the engineering staff needs to make sure the compiler can be used to develop software that can be sold without paying additional royalties. When adopting a technical standard, the engineering department needs to identify what licensing is needed to adopt the standard.

Patents are more complicated, mainly because there is a considerable time lag between filing for a patent and obtaining one, and because thousands of patents are filed. As a result, it is extremely difficult in the technology field to make sure a product does not infringe patents that are in the process of being issued. A classic strategy, therefore, is to file for patents on products in development with the goal of having some sort of protection. In the worst case, if another company accuses of patent infringement, having one's own patents enables one to counter-sue. In fact, this is an example of how an engineer can contribute directly to the competitiveness of a company – by aggressively filing invention disclosures and helping the company obtain patent protection on their innovations.

Commercial Aspects of Design

Another angle to consider in product design is financial. How does the engineering staff impact the economics of a product? There are two primary areas.

First, when the engineering team designs a product, they effectively dictate how much the product costs to manufacture. As you may recall, the list of components that are needed to assemble a complete unit is specified by engineering and is called the *bill of materials*. So the cost of the components is called the *bill of materials cost* or *BOM cost*. In most products, the BOM cost is the overwhelmingly dominant cost factor. Not only does the BOM cost need to be paid for every single product that is manufactured, but also the cost of manufacturing is normally substantially less due to automation (and perhaps due to the use of low-cost overseas labor, to be frank). Engineering dictates the BOM cost by completing the design, which dictates how many components are used and whether relatively expensive components are needed, and by qualifying the vendors for components, possibly choosing more expensive, higher-quality vendors. So there is, needless to say, tremendous pressure on engineering to reduce BOM costs.

The cost of manufacturing is called *conversion cost* (i.e., the cost of converting a bag of parts into a finished product). This cost includes labor, scrap, and everything else associated with manufacturing. This is also dictated by engineering, but the connection is more subtle. As discussed in the chapter on DFX, the art of product engineering lies in designing for manufacturability.

Products may have other costs, such as licensing and royalties, again dictated by the technologies selected by the engineering staff.

There are typically other direct costs, such as transportation or shipping of the manufactured product. All of these constituents of cost (BOM cost, conversion cost, royalties, shipping, etc.) are called *recurring expenditures* because they are paid for every single product assembled. They are also lumped together into the *cost of goods sold* or COGS.

The second element of engineering cost is in the effort required to carry out the design process. Most of the cost is in the salaries of engineers, but there may also be costs such as the construction of tooling. *Tooling* is the term for fixtures that are needed to manufacture custom components, such as chassis and custom integrated circuits.

There are quite a few non-engineering-related costs as well, such as the cost of advertising, sales commissions, dividends paid to shareholders, and the salaries of everyone else in the company.

These are called *nonrecurring expenditures* (or NREs). The finance department then engages in an accounting exercise to spread out the cost of the NRE's across the product while it is being manufactured, a process that includes *amortization* (spreading out the cost) and *depreciation* (acknowledgment of the fact that tooling is used up over time).

The company then turns around, adds some dollar amount, and charges its customers more for the product than it cost to manufacture. The amount that a customer pays for the product is called *price*. (Note the contrast – cost is what the company pays out and price is what is paid to the company.) The difference between customer payments and cost is called *profit* but that is not the end of the story.

The total of all the money paid to a company by its customers is called *revenue*. The difference between revenue and COGS is *gross profit*. If expressed as a ratio (revenue minus COGS divided by COGS), it is called *gross margin*. The difference between revenue and all costs (both COGS and amortized/depreciated NRE's) is called *net profit*.

Example 18.1

You are designing a product. You estimate the bill of materials cost to be $27.43, and the manufacturer estimates conversion costs at $3.75. Shipping per unit is $1.00 and licensing and royalties total $2.25 per unit.

(a) How much should the company charge if its gross margin target is 40%?

The total per-unit cost is $27.43 + $3.75 + $1.00 + 2.25 = 34.43. So for a price P we have $(P - 34.43)/P = 40\%$. Thus $0.6P = 34.43$ and so $P = \$57.38$.

(b) The company incurred additional costs of $465,000 associated with the product. How many units need to be sold to obtain a net profit of $100,000?

The profit per item is $57.38 - $27.43 = \$29.95$. So the company makes $29.95 on each unit it sells. To obtain a net profit of $100,000, there must be enough sales to cover the $465,000 costs and then make an additional $100,000. So the company must sell $565,000/\$29.95 = 18,865$ units.

There are standards in the accounting industry that delineate these distinctions, and there are (now) federal laws that mandate honest accounting, such as the *Sarbanes-Oxley Act* (or SOX). SOX has the interesting property that the CEO can go to jail if the accounting is not honest.

What does all this have to do with engineering? First, as noted above, much of the elements of product cost are dictated by engineering. Engineering can drive improvements in gross margin directly by manufacturing good products at lower cost, and can drive improvements in net margin by implementing more efficient design processes. Second, understanding "the rest of it" is a very important exercise for an engineering professional. It lets one understand how the company makes money, and lets one track the financial health of your company.

Engineering Notebook – Predicting What Comes Next

At one of the companies I worked for, I noticed that the company stopped investing significant money in research and development. Ongoing product development continued, but management became much less interested in new technologies and new product ideas. I was confused because the company had quite a bit of cash in the bank – over $1 billion – but did not seem to be interested in doing anything with the money. By not investing in research, the company was keeping the cash in the bank.

So what does it mean if your company has a big pile of cash on hand? For starters, it means the company is well-managed. But in finance, a pile of cash is considered a waste – it needs to be invested in some profitable activity and put to work, not allowed to lounge around in a bank somewhere.

If a company is sitting on its cash, it actually normally means the management wants to sell the company. Having a pile of cash makes the company more attractive for potential buyers. (After all, it makes the company look well-managed.)

The other interesting thing is what happens to the money – follow carefully here. Company "B" is buying company "A", and company "A" has a lot of cash in the bank. So Company B has to buy the cash. That is, they have to pay more for Company A by an amount equal to the cash. Where does the money go when Company A is bought? When a company is bought, what actually happens is that the purchasing company buys all of the stock of the company. Thus the money used to buy Company A goes to the shareholders of Company A, since the shareholders (the people who own stock in Company A) are the actual, legal owners of Company A. In other words, the money is legally transferred from Company B's bank account to the shareholders of Company A. And remember that Company B had to come up with extra money to buy Company A. So, by piling up cash and then selling Company A, Company A's management team is legally taking the cash used to buy the company and giving it out to the shareholders, the owners. One may complain about this or claim some sort of corruption, but one has to understand that the shareholders are the legal owners of the company, and therefore of the cash.

Sure enough, my company was bought shortly after I figured all of this out. The management team, which had a great deal of stock in the company, made out well. They were also all near retirement age, which explained a lot.

Project Management

Another angle of product design is the management of the design process itself. This is usually considered from the standpoint of the design schedule.

The process of preparing a schedule requires three steps, listing the steps needed to complete the design, estimating the time needed to complete each step, and then tracking the dependencies between steps. A common way of doing this is by preparing a *Gantt chart*.

For example, consider (an abbreviated version of) the design process from laying out a circuit board to assembling and testing units.

First, list the steps involved. The steps of the design include selecting components, drawing a schematic, laying out a circuit board, ordering test fixtures, ordering components, ordering the circuit board itself, assembling units, and testing them.

Second, determine how long each step takes. For example, assume that each of the selection and design steps takes two weeks, components take four weeks, and the circuit board takes one week to be fabricated. Assume that assembly takes one week and testing takes 2 weeks.

Third, list the dependencies. The steps of selecting components, drawing a schematic, and laying out the circuit board must occur in order. Ordering the components is possible after the components have been selected. Ordering the circuit board and the test fixtures can only occur after the board has been designed. The product can only be assembled after the components, circuit board, and test fixtures arrive at the manufacturing facility.

Armed with this information, a Gantt chart can be drawn (Fig 18.1).

The Gantt chart shows clearly that assembling units can only occur once the steps of ordering the parts, ordering the board, and ordering test fixtures are complete. Whichever one of the three takes the longest will dictate when the units can be assembled.

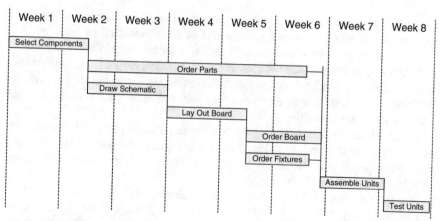

Fig. 18.1 Example of a Gantt chart

The path that takes the longest is called the *critical path*. This information serves two roles. First, it shows how long the project will take under the best case. This enables the staff to post a realistic schedule to the rest of the company. Second, it shows what happens if part of the schedule slips. For example, if part of the schedule is not on the critical path slips, the completion date will not change. If an early step needs to be redone (e.g., if a component turns out to be unavailable) the Gantt chart can show what the result will be on the schedule.

Even if, as a manager, you do not literally draw a Gantt chart, being able to draw a mental one is extremely helpful. It is very important to understand the critical path so that the urgency of each step of the design process can be communicated accurately. It also helps manage a project across multiple functional groups. For example, it enables a design manager to communicate effectively with the procurement department and the manufacturing staff.

There is specialized software that can be used to draw very elaborate Gantt charts, such as Microsoft Project, but the key is to perform just enough charting to be useful.

Risk Mitigation

So, armed with a detailed map of the design process and knowledge of design, what can go wrong? There turn out to be three types of design risk, and there are strategies that can minimize each type of risk.

The first is *technical risk*, the risk that the product will not meet its requirements. At first, there does not seem to be much anyone can do other than to do good design, but there are some very important steps to take. The first is to understand the types of design mistakes that are the most likely to be made and have design reviews that focus specifically on them. For example, making correct power and ground connections in a schematic is extremely tedious. Creating a mapping of signals to package pins is highly error-prone. These can be double-checked by hand.

Engineers should invite coworkers to review and double-check work frequently.

It is very important to start with the technically riskiest parts of a design. There is often a motivation to take shortcuts and do the easiest steps first, but starting with the riskiest parts enables the most amount of time to be put on the parts of the project that need the most amount of time.

Finally, once the manufacturing facility has made some prototypes, go through the units that do not work (also known as the *bonepile*). The bonepile is actually a valuable source of information – it shows you how the product is likely to break and likely to be manufactured incorrectly.

The second is *schedule risk*, the risk that the product will be late to market. One key to publishing realistic schedules is to add a small amount of extra time to each design step, time to make a few mistakes and catch back up. By adding extra time, the schedule is more in line with reality. By adding a small amount to each step, the schedule is harder for management to artificially compress. Remember that catching up on an aggressive schedule that is behind is nearly impossible.

There is an urban legend that Motorola was 6 months late with the 68000 processor and, as a result, IBM was forced to build the first PC using Intel processors. This may have been the most disastrous schedule slip in the history of computing. The lesson here is that schedule risk is often the worst because it opens up competition.

The third is *cost risk*, the risk that the product will cost too much to meet its goals (usually the margin target of the company versus the maximum price the company believes it can charge). One step is to build extra cost into the bill of materials as a contingency, some margin so that management does not come unglued later if costs go up. Working with the vendors to obtain better pricing is obviously essential, and performing DFM (design for manufacturability) can reduce conversion costs.

Engineering Careers

All of this talk of managing and projects leads to the next thought – have you thought about whether you want to be in engineering management one day? Or are you content to become a technical expert? For that matter, are you scared of being chained to a desk, or scared that you won't be?

This all leads to some thinking that you need to start now, before you graduate. You can and should take steps now to prepare for the rest of your career. For example, if you think you might want to be in management, it might be helpful to take an elective in leadership or technical management.

Once you graduate with an undergraduate degree, you have three basic career paths ahead of you. One is to become a technical expert, an uber-engineer, over time. In other words, you become and remain an engineer, eventually becoming the oldest engineer on staff and therefore the "go-to guy." A second path is to go into engineering management, usually after having been an engineer for a few years. The keys to being effective in this path are to be a good engineer (so you can, in turn, understand what your staff does every day) and to be good at thinking about the entire process (the Gantt chart in your head). A third path is to take your engineering degree and do something else with it. Some go into technical sales, a good path if you like engineering and have excellent people skills. Some get a masters in business administration (MBA) and move to up senior management by having a good grasp of both business and engineering. One path many are not aware of is law, of taking an engineering degree and getting a law degree. The law profession rewards lawyers who are also technical experts.

Fortunately, you don't have to know which one you want to do immediately, and going down one path is a reversible decision (as long as you reverse early enough). Keep your eyes open, talk to others who have made similar decisions, and stay well educated.

Appendix

1. Consider safety certification for a product to be sold in the US and in Europe.

   (a) Which US agency requires safety certification for products to be used in offices?
   (b) What type of laboratory is required for safety certification in the US?
   (c) Name a US and international standard that govern safety certification.
   (d) What marking is needed on components that are to be used in safe products?
   (e) What mark is needed for European markets?
   (f) For European markets, what testing is needed in addition to safety certification?

2. Define the following terms:

   (a) Bill of materials cost
   (b) Conversion cost
   (c) Nonrecurring expenditure
   (d) Recurring expenditure
   (e) Cost of goods sold
   (f) Tooling
   (g) Amortization
   (h) Depreciation
   (i) Cost
   (j) Price
   (k) Gross profit
   (l) Gross margin
   (m) Net profit

3. A product is being designed. The bill of materials cost is $25, and the conversion costs per product are $8. Royalties are $3 per unit, and there is a separate one-time patent-licensing cost of $40,000. The product requires $50,000 of tooling. Assume nonrecurring expenditures are amortized over 10,000 units.

   (a) What is the cost to manufacture one item? Consider per-unit cost both for the first 10,000 units and all units after that.
   (b) How much should the company charge if the gross margin target is 40%? Consider both the first 10,000 units and units after that.
   (c) How many units need to be sold to cover the nonrecurring expenses?

4. Consider the Gantt chart shown in Fig. 14.1. On a project you are managing there are the following times:

   – Select components: 1 week
   – Order (and receive) parts: 5 weeks
   – Draw schematic: 2 weeks
   – Layout board: 1 week
   – Order (and fabricate and receive) boards: 1 week
   – Order (and receive) fixtures: 1 week
   – Assemble and test units: 1 week

   (a) Construct a Gantt chart of the schedule.
   (b) On your Gantt chart, highlight the critical path in red.
   (c) How many weeks will the project take?
   (d) What will happen if the following changes occur:

      (i) Ordering parts delayed one week
      (ii) Circuit boards delayed one week

5. What are the three primary technical risks?

Index

© Springer Nature Switzerland AG 2022
S. H. Russ, *Signal Integrity*, https://doi.org/10.1007/978-3-030-86927-4

Printed in the United States
by Baker & Taylor Publisher Services